TEDDY ROOSEVELT & THE MARQUETTE LIBEL TRIAL

TEDDY ROOSEVELT & THE MARQUETTE LIBEL TRIAL

MIKEL B. CLASSEN

THE History PRESS

Published by The History Press
Charleston, SC 29403
www.historypress.net

First published 2015

Manufactured in the United States

ISBN 978.1.62619.557.8

Library of Congress Control Number: 2014953418

To Dr. John Watanen, who told me I should be a writer; Dr. Thomas Hruska for making sure I could; Dr. Jon Saari for showing me the tools of history research; and Dr. Gerald Waite for the local history/journalism assignment that led to this book.

Contents

Acknowledgements

Many thanks to Rosemary Michelin at the Marquette Regional History Center for her help with the research and acquiring some of the very special photos in this book and Jack Deo from Superior View Studios for helping me with his great collection of historical Roosevelt photos. And thanks to Mary L. Underwood for additional background research and help with this project in countless ways.

Introduction

I first learned about this story when I was at Northern Michigan University in Marquette, Michigan. One of my professors assigned me to write a paper on an aspect of local history that was associated with journalism. This story was the result of that assignment.

That was thirty-five years ago.

Now, in revisiting the subject for this book, I've been able to look at this piece of history in much more detail. I have the opportunity to try to reflect on some of the fascinating stories that were a part of the libel trial testimony and the incidents surrounding it. While doing the research at the Marquette Regional History Center, I was shown some of the many photos they had on hand, which included a scrapbook from wildlife photographer and naturalist George Shiras 3d. In this book were several photos of Roosevelt during the trial at the Shiras house. The pictures weren't credited in the scrapbook, but I strongly feel that many of these pictures were shot by Shiras himself and have never been published before.

I have left Teddy Roosevelt's testimony and that of George Newett complete and unabridged. I felt this helped preserve some of the drama of the trial, but it also provides an in-depth look at Theodore Roosevelt in every aspect of his life from the extraordinary to the mundane.

The testimony of the trial expresses how important it was to Roosevelt to clear his name of the gossip of alcoholism that surrounded him. He had a legitimate fear that the rumors of drinking would follow him into history as a part of his presidential legacy. Former presidents like Grant and Jackson had tales of their drinking spoken of more than their accomplishments. This was not a stain Roosevelt was willing to wear into history.

There may have been something else as well, something deeper, something more personal that motivated Roosevelt to so deeply resent being labeled "a drunk, a boozer and an abuser of alcohol." Teddy Roosevelt had witnessed firsthand the devastation alcoholism can wreak.

Teddy Roosevelt (Theodore Roosevelt Jr., known as "T.R." to his family) was the oldest son of glass businessman and philanthropist Theodore "Thee" Roosevelt Sr. and socialite Martha Stewart "Mittie" Bulloch. He was one of four children, having an older sister, Anna, and a younger brother and sister, Elliot and Corinne. The Roosevelt children came from a stable, loving home. Thee and Mittie adored their children, and it was one of their greatest fears that T.R., who was born sickly and asthmatic, would not survive. They also feared that, in the event that he did live to see adulthood, their sickly son would never prosper in the world.

Teddy's youth was, in large part, shaped by his poor health and the family's desperate fight to overcome his severe asthma and the debilitating impact it had on his body and personality. He experienced recurring sudden nighttime asthma attacks that caused near-deathlike experiences of being smothered to death, terrifying the boy and his parents. Doctors had no cure, and his frantic parents would bundle him into their carriage and whip the horses through the streets of New York at a breakneck speed in a desperate attempt to force air into the child's lungs.

Thee Roosevelt had a gym constructed in their home so that his son could work at strengthening his sickly body. "Your body will betray you," he warned the boy. So Teddy exercised.

Much of Teddy's early schooling was accomplished at home due to his poor health, and Teddy struggled to keep up academically with other children. Thee Roosevelt, worried that his son would never catch up either academically or physically with other children, set aside a small trust fund to support Teddy in the event that Teddy was never able to support himself. "I have given you your bread and butter," the senior Roosevelt said, "but if you want toast and jam, you must work for it yourself." And encouraged by his loving parents, Teddy studied hard and worked to become a man they would be proud of.

However, not all of the Roosevelt children had to struggle to succeed. Mittie and Thee Roosevelt had another son, and he was everything young Teddy was not. Elliot Roosevelt was a picture of the perfect child: strong, robust, amazingly attractive and charming. He was quick to excel at everything he tried. Young Elliot was good at sports, an excellent student and remarkably personable. His parents were certain that Elliot would excel at whatever he chose to do. When Elliot went away to university, he did well without trying and spent a great deal of his time drinking and carousing with his classmates. Elliot, claiming health reasons, left the university before graduating. This may have been the result of his early "casual drinking problem."

Handsome, charming Elliot continued with his drinking. He married Anna Rebecca Hall, and they had three children, Anna Eleanor Roosevelt (called Eleanor, who was later to marry a distant cousin, Franklin D. Roosevelt), Elliot Roosevelt Jr. and Gracie Hall Roosevelt (called Hall by the family).

Although Elliot adored his children, family life did not deter him from drinking, and his drinking had dire consequences. Due to his drinking problem, Elliott was exiled to a sanatorium in Abingdon, Virginia, for a program we would refer to as rehab today. On occasion, Elliot would, to the joy of his children, return home for a few days. But he would always go back to drinking, and that would send him back to Virginia.

Theodore Roosevelt became the conservator for his beloved, brilliant, alcoholic brother.

Elliot Roosevelt attempted suicide in 1894 at the age of thirty-four. Although some would argue that the attempt was a failure, Elliot died of complications a few days later. Eleanor was ten, and young Hall was only three years old. Teddy, who had lost the brother he loved, took responsibility for Elliot's children, walking Eleanor down the aisle when she married Franklin.

Sadly, this wasn't the last time Teddy had to deal with the crushing pain of alcoholism. Unfortunately, his beloved niece Hall followed in her father's footsteps.

So Teddy Roosevelt may have had more than one reason for resenting the label of "boozer." Teddy understood all too well what that title entailed.

1

Of Campaigns, Assassins and the "Roosevelt Way"

It all began in 1912, the year Theodore Roosevelt made his run for president as an independent candidate. Roosevelt was the candidate of the Progressive Party, also known as the "Bull Moose" Party. The campaign made its way across America, hitting town after town, none too big or too small. Eventually, the campaign train made its way to Michigan's Upper Peninsula, including the cities of Marquette and Ishpeming. There were others along the way, but it was here that an important chapter in Roosevelt's life would be written. At least, it was important to him.

The campaign trail in those days did not include microphones. A crowd had to be reached through sheer volume of voice, and of this Roosevelt was a champion. He was well known for his boisterous and animated speeches. His enthusiasm and volume displayed during his speeches had often led many to speculate on his sobriety during these moments of passion. It was rumored by some, mostly those who disliked Roosevelt, that his "bully" came from a bottle.

During a time when vices were common and the health hazards of those lifestyles were unknown, Theodore Roosevelt led a very healthy lifestyle. As an athlete, military man, outdoorsman and explorer, physical fitness had always been something he maintained to the best of his ability. Smoking and drinking weren't things that appealed to him.

As president, matters of state and protocol occasionally demanded a toast of champagne or wine, in which he would participate, but his personality of volume and passion would at times be misinterpreted as excessive drinking.

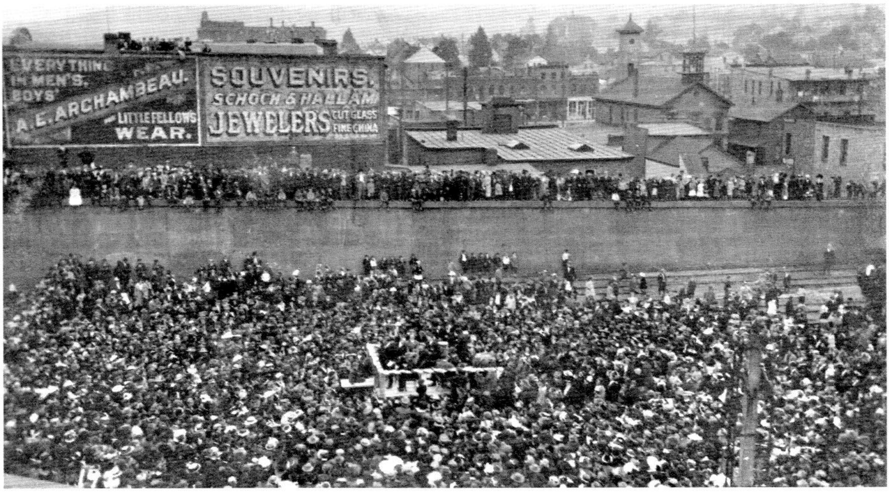

Teddy Roosevelt's campaign train rolled into Marquette to a huge crowd in 1912. Roosevelt had considerable support as an independent candidate, particularly in Michigan's Upper Peninsula. *Courtesy of Jack Deo of Superior View Studio.*

Rumors cropped up everywhere, and new stories of his overindulgence would follow virtually every appearance. It got to the point where all Roosevelt had to do was be present and new rumors would start.

A story told by newspaper reporter Jay Hayden, who was following Roosevelt for the *Detroit News* during the 1912 campaign, illustrates the extent of the problem:

> *I was told to board a train en route to the* [Republican] *convention at Toledo and find Roosevelt, who was reported aboard. Peeking into various compartments, I found the Colonel alone and reading. We had a sparkling half-hour of warm conversation before other visitors intruded.*
>
> *I scarcely had stepped from the train in Chicago before I began hearing that Roosevelt had been roaring drunk on the trip and had smashed dishes in the dining car.*

Prohibition was becoming a hot issue during the campaign. Roosevelt had claimed that his party "was a party that would dodge no issue." Literally a few seconds after he made that claim, a small, gray-haired man climbed up on a chair and asked, "Mr. Roosevelt, where do you stand on prohibition?" Roosevelt windmilled his arms and made some noises, but no one actually understood the answer. It was one of the few times when he was completely unintelligible. The subject of prohibition was one he preferred to avoid.

This did not help dispel the rumors of drinking; in fact, it was gasoline on the fire. His political opponents reveled in all of this and encouraged the rumormongering. There was considerable animosity toward Roosevelt, including from many of his former supporters. When he had lost the nomination for president by the Republican Party to William Howard Taft, he broke from the party and accepted the nomination from the Progressive Party. This move split the Republican Party and virtually ensured the Democratic nominee Woodrow Wilson's victory.

The rumors of his drinking made Roosevelt angry. He did not want his legacy tainted with the perception that all he had accomplished had been the actions of a drunk. Most importantly, in Roosevelt's eyes, they weren't true. "If this slander is ever printed in so many words, I will bring suit for damages and settle it once and for all," the colonel warned during the campaign. Teddy had had enough.

When Roosevelt's campaign train pulled into Marquette, Michigan, it was met by a huge crowd. It was a city-wide party, with Roosevelt laughing and joking with those in the crowd. A makeshift stage had been built in the railroad yard that ran through the heart of the city. The city residents gathered in close to hear the former president and now-independent candidate.

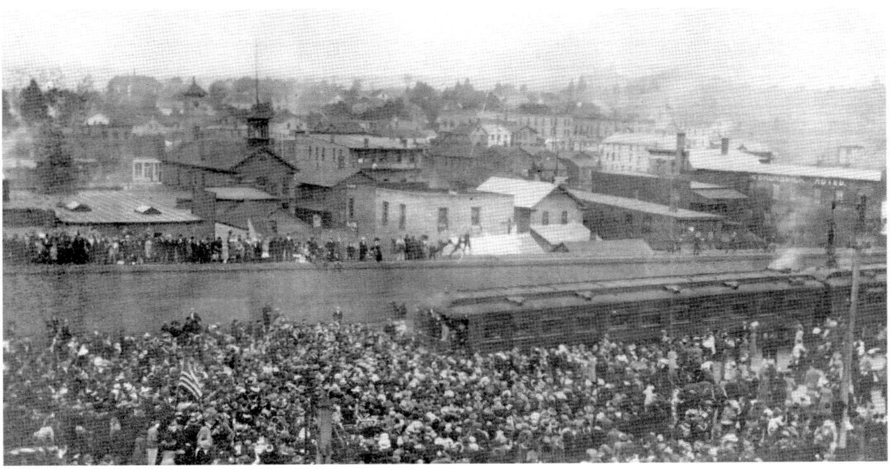

A view of the immense crowd that filled every available space for a view of the former president. Roosevelt's popularity is evident in this picture. *Courtesy of Marquette Regional History Center.*

During his speech in Marquette, Roosevelt spoke about the issues of the day, focusing on what he saw were the weaknesses and corruption of some of his former party mates in Michigan. Specifically, he attacked what he called the "steel trust," which he blamed for taking over the Republican National Convention the previous year in Chicago. T.R.'s more radical antitrust views were quickly making him an adversary of the industrialists who united against him. They had resulted in Roosevelt losing the Republican nomination, something he was bitter about. On the campaign trail, he spoke against the industrialists with every opportunity.

With Marquette in the heart of the Michigan Iron Range, Roosevelt took full advantage. He singled out local individuals whom he felt were in the pocket of the steel trust. "The steel trust is here in Marquette County, and its attorney, the congressman against whom—"

In the middle of the accusation, Roosevelt was interrupted by a Marquette man, Mr. John Van Evera: "That is not true." Challenging Roosevelt on his own "bully pulpit" was a bad idea. Van Evera was completely outclassed in debate.

Roosevelt, without hesitation, threw back, "You stand for theft, and you stand for lying and false witness bearing. Another thing I will give a chance to deny, that every paper influenced by the steel corporation in Marquette and by the standpatters is against us in this county."

Van Evera was not deterred: "I'm not afraid of a Bull Moose."

Roosevelt never missed a beat, responding, "It is perfectly natural that you should object to hearing the truth told about the side you are championing; and it is perfectly natural that you should come here to try to interrupt a meeting in which I am exposing the falsities and misinterpretations of your side."

"Then tell the truth," continued Van Evera, undaunted. It would be his last comment. He couldn't have provided the former president a better opportunity to blast the local Republicans.

"Mr. Young [local Congressman Horace O. Young] is the ex-attorney of the Steel Trust, and his law partner is attorney for the Steel Trust now," continued Roosevelt. With no more interruptions, he went on:

> *I understand, sir, that I am telling you the truth; I speak here from the information given me; but when I speak of the Chicago convention of last June, I speak of what I know. You are supporting the receivers of stolen goods, and a man engaged with the theft; and if you are a man of intelligence and education, you are acting as dishonorably as if you were supporting a man who had stolen a purse. Now you ask to hear the truth.*

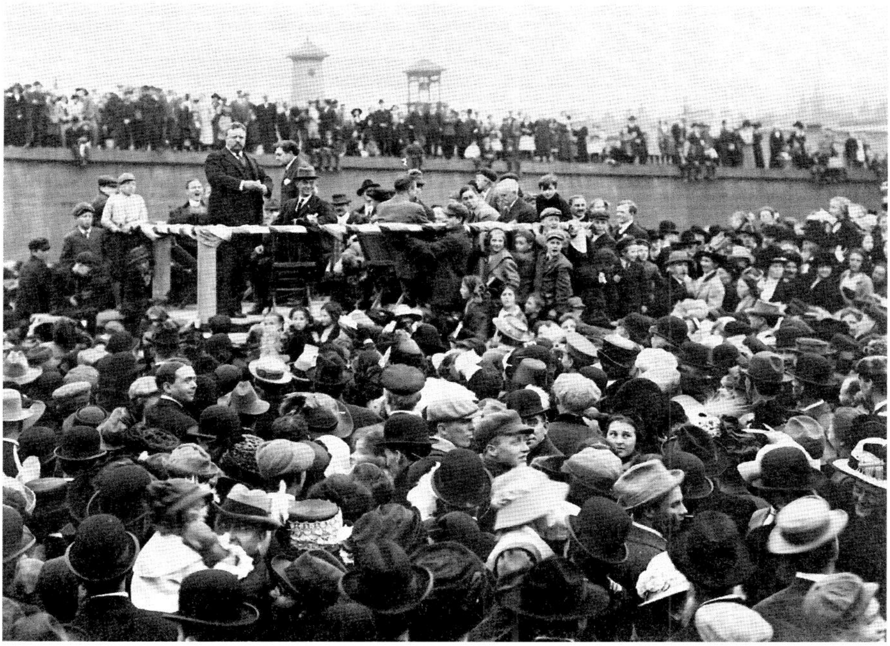

Roosevelt gives his speech to an attentive crowd. He can be seen pointing to emphasize a political point. It was this animated and boisterous style that led to the rumors of his rampant drinking and overindulgence, rumors kept alive by political opponents hoping to discredit his reputation and tarnish his image. *Courtesy of Marquette Regional History Center.*

> *You have heard it. A man who approves of the commission of theft, or who brazenly defends it, is no better than the thief himself.*

Roosevelt's speech was a success. Local headlines heralded his campaign tour: "'Big Bull Moose's' Tour a Continual Triumph" and "Upper Peninsula Turned Out More Than 40,000 People Wednesday to Welcome the Great Progressive Chieftain." Another read, "He Was Seen and Heard in Marquette County by Larger Throngs Than Ever Before Had Greeted a Great National Leader." It seemed like more success for the former statesman.

In the audience listening to Roosevelt was a staunch Republican by the name of George Newett. He was the publisher of a local newspaper called the *Iron Ore*, which had its offices in Ishpeming. Newett was angry. He felt betrayed by Roosevelt and went home and penned a scathing editorial titled "The Roosevelt Way." It went too far.

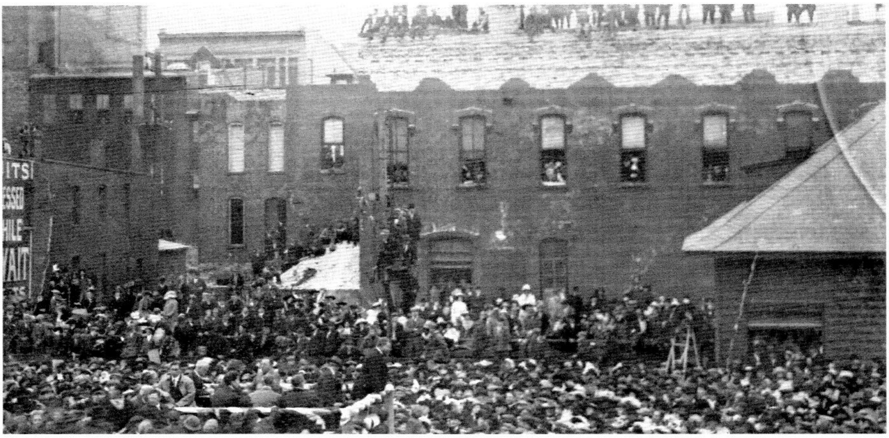

Roosevelt gives an enthusiastic speech at his 1912 campaign stop in Marquette as part of his Upper Peninsula tour. People on the roofs, people in the windows—anywhere anyone could get a vantage point, someone filled it. *Courtesy of Jack Deo of Superior View Studio.*

Jay Hayden, the *Detroit News* reporter, picked up the story:

> *Two days later on Oct. 14, Roosevelt was handed a copy of the editorial by his party's secretary, Oscar King Davis. The ex-president was suffering from a sore throat from his marathon speech making in his Chicago headquarters. He read Newett's attack: "Roosevelt lies and curses in a most disgusting way; he gets drunk, too, and that not infrequently, and all his intimates know it." He whispered to Davis, "Let's go at him!"*
>
> *While en route to Milwaukee where he was to speak that night, Roosevelt sent instructions to Henry M. Wallace, Progressive national committeeman from Michigan. Wallace was to retain a lawyer and file a libel suit against Newett.*

Milwaukee turned out to be a fateful place for Roosevelt. On October 14, Roosevelt was picked up at his hotel on the way to a speech. He stepped into the back of the waiting car, sat down and then rose to acknowledge the crowd, waving to the cheers of countless residents waiting for a glimpse of the colonel. A man standing seven feet away drew a revolver from his vest and fired, striking Roosevelt in the chest. The shot knocked him back into his seat.

The man who pulled the trigger was John Schrank, a New Yorker who claimed he was following the orders of McKinley's ghost. The story goes that Schrank was writing a poem in the night when the ghost of McKinley appeared to him and told him to kill Roosevelt and deny him his third term. Apparently McKinley's spirit was very persuasive.

When examined by one of his personal doctors, Dr. Terrell, it was discovered that the bullet had gone through Roosevelt's steel glasses case and the manuscript of his speech, which was in his pocket and was quite thick. They had slowed down the bullet so that nothing vital was hit, but it was lodged in his rib. Roosevelt insisted on completing his speech before going to a hospital.

He stayed in the hospital for several days, not only convalescing, but also resting from his whirlwind campaign tour. In the Upper Peninsula, he had made two speeches before breakfast, twenty-six more before dinner and two after dinner. That many daily speeches was typical for Roosevelt. The colonel was exhausted.

For many years it was believed that it was here, while he was recuperating, that Roosevelt was handed the issue of the *Iron Ore* containing Newett's editorial. Though it became the common story, it was not true.

In May 1913, the suit that Roosevelt had filed came up for trial, and Roosevelt returned to Marquette. In the beginning, the case looked like it was open and shut for Roosevelt, but as the trial approached, a private investigator hired by Newett had acquired no fewer than forty depositions, all swearing to have seen T.R. in a state of intoxication. Roosevelt prepared himself for the first time for the possibility that he could lose. He commented as he left for the trial, "I did not believe I had any new depth of human nature to plumb, but I was really unprepared for perjury on a gigantic scale, perjury by scores of alleged witnesses. I am really puzzled by the extreme malignancy of the hatred which purchases such testimony or offers it without purchase."

2

George Newett and the *Iron Ore* Newspaper

George A. Newett was feeling confident with his position leading up to the trial. "*Iron Ore* has long held a pretty good reputation for sticking to the facts," he said. The suit, in his mind, was "another desperate effort to bolster up a fast declining cause." They were fiery words. Newett plainly thought he had Roosevelt right where he wanted him.

Very little is known about Newett and the newspaper that drew Roosevelt's ire. George Newett founded the *Iron Ore* in 1872 when he moved to Ishpeming from Janesville, Wisconsin. He was an unflappable supporter of the Republican Party and was a teetotaler in his personal life. Newett was married and had two sons, W.H. Newett and George A. Newett Jr. The pair took over the paper when George Sr. died in 1928. Ironically, Newett was appointed postmaster of Ishpeming by Theodore Roosevelt.

The *Iron Ore* newspaper reported on the mining news of the day and flourished through the early days of the iron-mining boom. It became an arm of the mining industry and was popular within the iron and steel community. It boasted subscribers throughout the world. A copy could be found in the offices of every mine on the Marquette iron range as well as on other iron-ore ranges in Michigan and Minnesota. If it had to do with mining, the story could be found in the *Iron Ore.*

As a part of the mining industry, Roosevelt's continuing radical stance on a better life for the workingman was something industrialists were against. A fair day's wage for a fair day's work was an expensive prospect for business in general. The steel industry, with its massive employment base, wanted

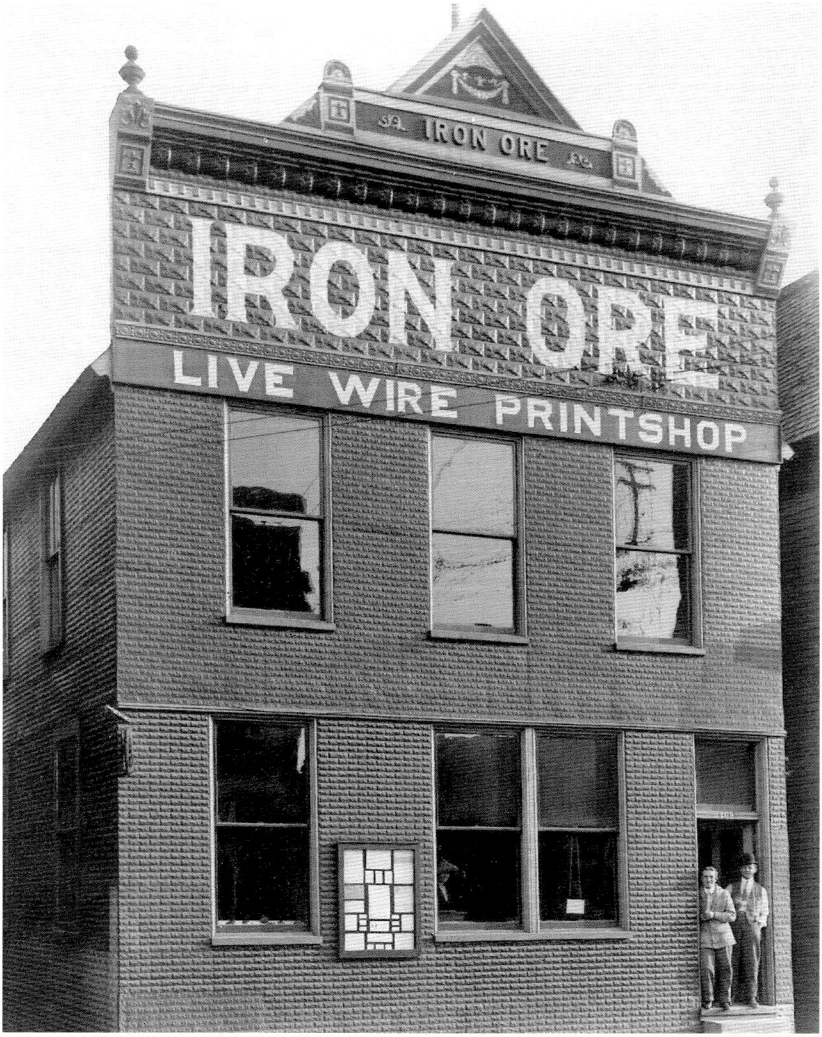

The offices of the Ishpeming *Iron Ore* newspaper. It was here that George Newett ran his Republican-oriented newspaper and printed his now-famous editorial "The Roosevelt Way," accusing Roosevelt of excessive drinking and rude behavior. *Courtesy of Jack Deo of Superior View Studio.*

nothing to do with Roosevelt's new ideas. Hence, Roosevelt's perceived coup at the Republican National Convention in Chicago. William Howard Taft was chosen even though Roosevelt had won every primary in the country except one.

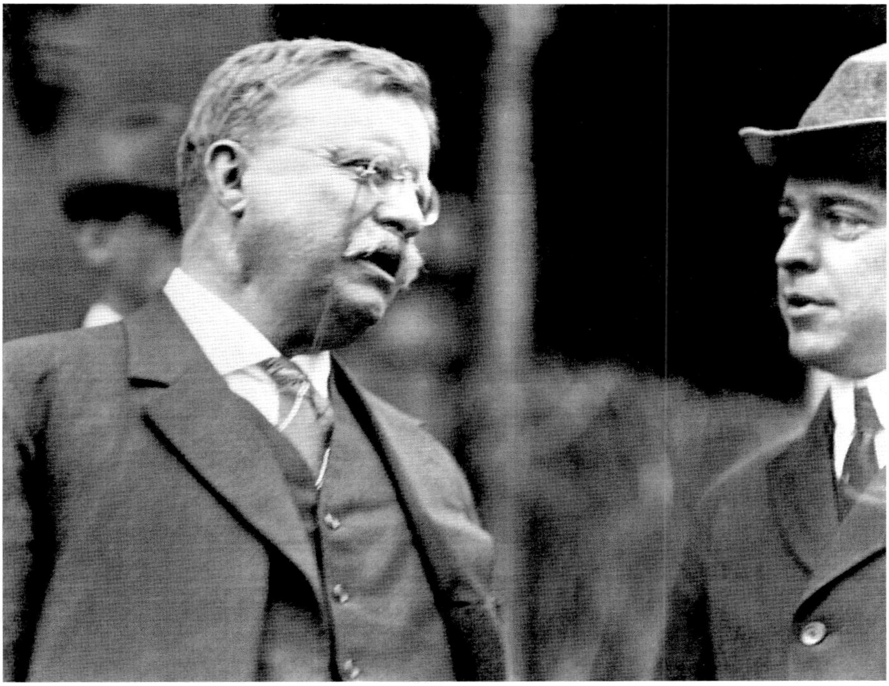

Roosevelt and Joseph G. Cannon in Marquette. Representative Cannon, or Uncle Joe as he was called by locals, played host to Roosevelt at Cannon's seventieth birthday party at which Roosevelt led a toast of champagne. This resulted in one of the accusations of drunkenness. *Courtesy of Jack Deo of Superior View Studio.*

George Newett was a staunch Republican, and the editorials in the *Iron Ore* reflected his party affiliation. When Roosevelt went over to the Progressive "Bull Moose" Party, Newett remained a loyal Republican and did not split off to support T.R. like many did.

When Roosevelt made his public denunciations of Representative Horace Olin Young, a close friend of Newett's, Newett was angry. He heard from a man named James Martin Miller, who swore he had seen Roosevelt drunk at the seventieth birthday party for Representative Joseph G. "Uncle Joe" Cannon in 1911. Roosevelt had led a toast of champagne honoring Uncle Joe's birthday. Miller swore out a deposition saying Roosevelt was drunk and swearing loudly.

Consequently, after Roosevelt's campaign stop in Marquette, Newett went back to his newspaper and penned a scathing editorial. The editorial, "The Roosevelt Way," is printed on the following pages in its entirety.

The Roosevelt Way

According to Roosevelt, he is the only man who can call others liars, rascals and thieves, terms he applies to Republicans generally.

All that Roosevelt has gained politically he received from the hands of the Republican Party.

Had he won in the Republican convention in Chicago, then the Republican Party would still be a good party, and all others would have been made up of liars and thieves and scoundrels generally.

But if anyone calls Roosevelt a liar he raves and roars and takes on in an awful way, and yet Roosevelt is a pretty good liar himself. Where a lie will serve to advance his position, he employs it.

Roosevelt lies and curses in the most disgusting way; he gets drunk, too, and that not infrequently, and all of his intimates know about it.

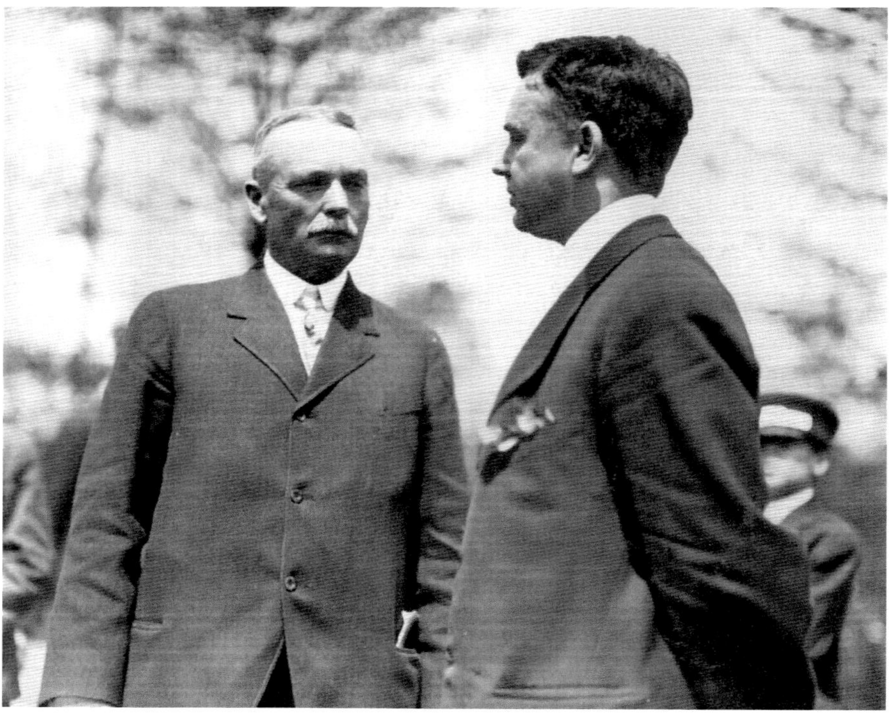

Pictured is George A. Newett (left) standing with an attorney named Fairchilds (right). This photo is from the Shiras scrapbook. *Courtesy of the Marquette Regional History Center.*

What's the use mincing things with him when he maltreats everyone not for him?

Because he has been president gives him no privileges above other men and his conduct is just as deserving censure as is that of any other offender against decency.

How can Roosevelt expect to go unlashed when he maliciously and untruthfully strikes out at other people?

It's just as Mr. Harlan said, he's the greatest little fighter in the country when he's alone in the ring, but he acts like a madman if anyone dares criticize him. He must do all the swearing and abusing of people. No one must question him. All who oppose him are wreckers of the country, liars, knaves and undesirables.

He alone is pure and entitled to a halo. Rats! For so great a fighter, self-styled, he's the poorest loser we ever knew.

Though many editorials attacked Roosevelt during his day, none had ever openly made the accusations Newett made. Newett had unwittingly set himself up to be the platform for Roosevelt to quell these rumors once and for all.

Newett wasn't intimidated. He was certain he was right and stood behind his statements. He went to everyone who had claimed knowledge of the former president's overindulgence and had them swear affidavits. All totaled, he collected forty of them. The stage was set.

3
Teddy Brings His "Big Stick" to Marquette

Theodore Roosevelt's lawyer, James H. Pound, a respected legal counsel from Detroit, filed an extensive Declaration of Intention on October 25, 1912, in Marquette County. It read in part:

> *Plaintiff avers that his reputation in every aspect is broadly and widely known throughout every part of the United States of America, as well as throughout the entire civilized world. And plaintiff avers that he has, as such, always behaved and conducted himself as a becoming, upright, self-esteemed, sober, temperate and abstemious citizen should, especially in the use of alcoholic and spirituous liquors. And that until the committing of the several grievances by the said defendant, as hereinafter mentioned, was always reputed esteemed and accepted by and amongst all his neighbors, acquaintances, friends and other good and worthy citizens of this union and whom he was anywise known, to throughout the civilized world to be a person of good name, fame and credit and of sober habits, to wit, at the city of New York, the city of Detroit, Michigan, also the County of Marquette, and elsewhere, wherever this plaintiff was in any wise known throughout the United States; and whereas also, the said plaintiff hath not ever been guilty, or until the time of the committing of the said several grievances, by the said defendant, as to have been guilty of the crime of habitual drunkenness or been suspected of having been a drunkard, or that he ever drank at all to excess, also of the offenses and misconducts here, now, mentioned and hereinafter set out, as having been charged upon him,*

> *plaintiff, by the defendant, George A. Newett, or of any other such offenses or misconduct.*

With this statement, a legal battle was initiated that would forever change the perception of an American president. Teddy Roosevelt knew his legacy was at stake and that the general citizenry believed the rumors of his overindulgence. If he prevailed, it could never be spoken of again.

On October 19, four days later, Attorney Pound filed a formal and detailed complaint:

> *That the said defendent, George A. Newett, did upon October, the twelfth, A.D. 1912, publish the following false, scandalous, malicious and defamatory words…"The Roosevelt Way."*
>
> *That the entire article is libelous. But that Theodore Roosevelt waives all claims for damages for any of the libels contained in said article, except the words, "Roosevelt lies and curses in a most disgusting way. He gets drunk, too, and that not infrequently and all of his intimates know about it."*
>
> *That Theodore Roosevelt does hereby begin an action of Trespass, in the Circuit Court for the County of Marquette and claims as his damages, the sum of Ten Thousand Dollars.*

Newett was officially challenged. Undeterred, he hired his own lawyer, William P. Belden, of his hometown, Ishpeming. They in turn filed their own response to Roosevelt's allegations of libel:

> *Take notice, the defendant will give in evidence and insist in his defense that the words charged in the plaintiff's declaration, were published in good faith, without any malice, and under circumstances creating a qualified privilege, vis.: That at the time the plaintiff was a candidate for the office of the President of the United States, and that as such candidate his public conduct and his fitness for said high office were properly subject to discussion as matters of common and general interest.*
>
> *And the said defendant will further give in evidence and insist in his defense, the plaintiff had been and was guilty of the facts and acts charged and imputed to him in the publication.*

Newett truly felt that what had been written was the absolute truth, and he set out to prove it. Newett and Attorney Belden began to search for anyone

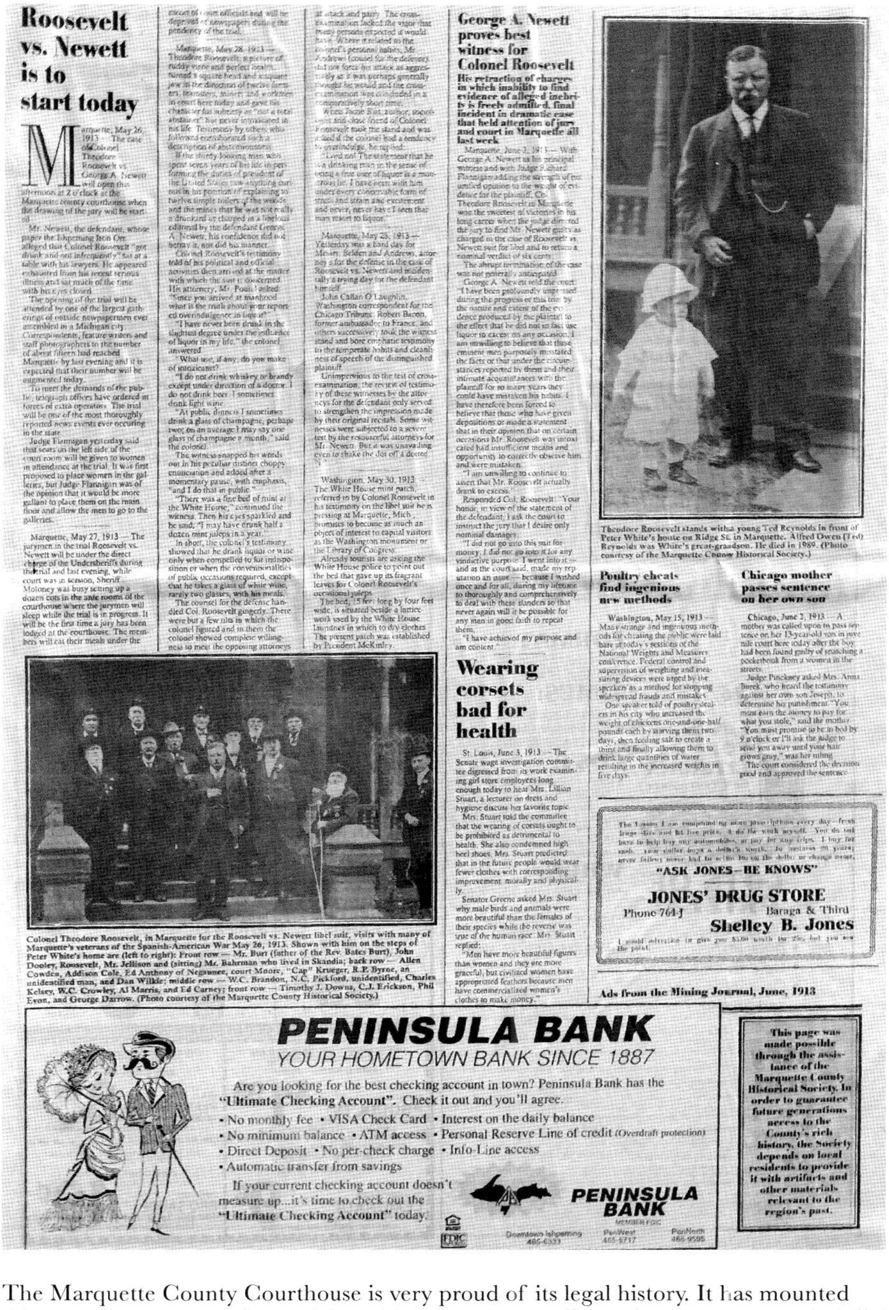

Roosevelt vs. Newett is to start today

Marquette, May 26, 1913 — The case of Colonel Theodore Roosevelt vs. George A. Newett will open this afternoon at 2 o'clock at the Marquette county courthouse when the drawing of the jury will be started.

Mr. Newett, the defendant, whose paper the Ishpeming Iron Ore alleged that Colonel Roosevelt "got drunk and not infrequently" sat at a table with his lawyers. He appeared exhausted from his recent serious illness and sat much of the time with his eyes closed.

The opening of the trial will be attended by one of the largest gatherings of outside newspapermen ever assembled in a Michigan city. Correspondents, feature writers and staff photographers to the number of about fifteen had reached Marquette by last evening and it is expected that their number will be augmented today.

To meet the demands of the public, telegraph offices have ordered in forces of extra operators. The trial will be one of the most thoroughly reported news events ever occurring in the state.

...

Marquette, May 28, 1913 — Theodore Roosevelt, a picture of ruddy vigor and perfect health, turned a square head and a square jaw in the direction of twelve farmers, teamsters, miners and workmen in court here today and gave his character for sobriety as "not a total abstainer" but never intoxicated in his life. Testimony by others who followed corroborated such a description of abstemiousness.

If the sturdy-looking man who spent seven years of his life in performing the duties of president of the United States saw anything curious in his position of explaining to twelve simple toilers of the woods and the mines that he was not really a drunkard as charged in a libelous editorial by the defendant George A. Newett, his confidence did not betray it, nor did his manner.

Colonel Roosevelt's testimony told of his political and official activities then arrived at the matter with which the suit is concerned. His attorney, Mr. Pound asked: "Since you arrived at manhood what is the truth about your reported overindulgence in liquor?"

"I have never been drunk in the slightest degree under the influence of liquor in my life," the colonel answered.

"What use, if any, do you make of intoxicants?"

"I do not drink whiskey or brandy except under direction of a doctor. I do not drink beer. I sometimes drink light wine.

"At public dinners I sometimes drink a glass of champagne, perhaps ...

...

George A. Newett proves best witness for Colonel Roosevelt

His retraction of charges in which inability to find evidence of alleged inebriety is freely admitted, final incident in dramatic case that held attention of jury and court in Marquette all last week

Marquette, June 2, 1913 — With George A. Newett as his principal witness and with Judge Richard Flannigan adding the strength of his unified opinion to the weight of evidence for the plaintiff, Col. Theodore Roosevelt in Marquette won the sweetest of victories in his long career when the judge directed ...

...

Theodore Roosevelt stands with a young Ted Reynolds in front of Peter White's house on Ridge St. in Marquette. Alfred Owen (Ted) Reynolds was White's great-grandson. He died in 1989. (Photo courtesy of the Marquette County Historical Society.)

Poultry cheats find ingenious new methods

Washington, May 15, 1913 — Many strange and ingenious methods for cheating the public were laid bare at today's sessions of the National Weights and Measures conference. Federal control and supervision of weighing and measuring devices were urged by the speakers as a method for stopping widespread frauds and mistakes.

One speaker told of poultry dealers in his city who increased the weight of chickens one-and-one-half pounds each by starving them two days, then feeding salt to create a thirst and finally allowing them to drink large quantities of water resulting in the increased weights in five days.

Chicago mother passes sentence on her own son

Chicago, June 2, 1913 — A mother was called upon to pass sentence on her 13-year-old son in juvenile court here today after the boy had been found guilty of snatching a pocketbook from a woman in the streets.

Judge Pinckney asked Mrs. Anna Burek, who heard the testimony against her own son Joseph, to determine his punishment. "You must earn the money to pay for what you stole," said the mother. "You must promise to be in bed by 9 o'clock or I'll ask the judge to send you away until your hair grows gray," was her ruling.

The court considered the decision good and approved the sentence.

Wearing corsets bad for health

...

Colonel Theodore Roosevelt, in Marquette for the Roosevelt vs. Newett libel suit, visits with many of Marquette's veterans of the Spanish-American War May 26, 1913. Shown with him on the steps of Peter White's home are (left to right): Front row — Mr. Burt (father of the Rev. Bates Burt), John Dooley, Roosevelt, Mr. Jellison and (sitting) Mr. Bahrman who lived in Skandia; back row — Allen Cowden, Addison Cole, Ed Anthony of Negaunee, court Moore, "Cap" Krueger, R.P. Byrne, an unidentified man, and Dan Wilkie; middle row — W.C. Brandon, N.C. Pickford, unidentified, Charles Kelsey, W.C. Crowley, Al Marris, and Ed Carney; front row — Timothy J. Downs, C.J. Erickson, Phil Evon, and George Darrow. (Photo courtesy of the Marquette County Historical Society.)

Ads from the Mining Journal, June, 1913

This page was made possible through the assistance of the Marquette County Historical Society. In order to guarantee future generations access to the County's rich history, the Society depends on local residents to provide it with artifacts and other materials relevant to the region's past.

The Marquette County Courthouse is very proud of its legal history. It has mounted this newspaper article of the trial, and it is kept in the offices of the court. *Photo by Mikel B. Classen.*

who was willing to testify against Roosevelt. It wasn't an easy task. But they did find some and began to collect depositions.

Roosevelt and his legal staff, knowing of the depositions, made a motion that all statements collected by Newett be suppressed until the trial. This way, any speculation and scandal would be kept out of the press until T.R. had his day in court. On February 10, 1913, Honorable Judge Richard C. Flannigan, who would preside over the trial, granted Roosevelt the motion and ordered all depositions to be suppressed. Roosevelt had won the first small victory.

In the meantime, Newett went forward with collecting witnesses and depositions. All of the witnesses swore that they had seen Roosevelt drunk on one occasion or another. One in particular, a James Martin Miller, had sworn he saw Roosevelt drunk and swearing at the seventieth birthday party for Congressman Joseph Cannon. He was going to be one of Newett's most important witnesses.

Attorney Pound was also out collecting witnesses and depositions. These witnesses would swear to the sobriety of the former president. Pound had heard through rumor that Newett was having trouble gathering witnesses. The word was that he had only a handful, none of whom were prominent citizens. Pound also heard that Newett's witnesses wanted to remain anonymous. Pound had gathered thirty-five witnesses, all of whom were anything but anonymous. Pound hoped to overwhelm Newett and bury him under a string of witnesses that he'd never be able to compete with, individuals whose credibility and integrity were beyond reproach. It became obvious that Roosevelt had gained the respect and loyalty of some of the greatest men in America for the era.

On May 25, 1913, that fact was fully impressed on the citizenry of Marquette County. Arriving on a special train, the Clearwater from Chicago, Roosevelt disembarked in Marquette in true Teddy Roosevelt fashion. He was jubilant, waving and immediately shouting to friends and those who'd come to see him from the city, including one friend in particular, George Shiras III or 3^{d}, as he had become known. It would be Shiras with whom Roosevelt would stay during the trial.

Along with Roosevelt came an entourage, an overwhelming group of individuals. With him was James R. Garfield, former interior secretary and son of President Garfield, wearing a straw hat; Robert Bacon, former assistant secretary of state; Truman H. Newberry, a former navy secretary; Gifford Pinchot, former governor of Pennsylvania; George B. Cortelyou, who had held three cabinet posts under Roosevelt; and famous journalist and Roosevelt confidant Jacob A Riis.

Also accompanying Roosevelt were several well-known newspapermen, two former Secret Service men, his personal physician, his valet and a pair of close relatives: nephew W. Emlen Roosevelt and cousin Philip Roosevelt. They had all come to testify, having intimate knowledge of Roosevelt's personal life.

A commentary in the *New York World* describes it like this:

> *The newspaper which Mr. Roosevelt is prosecuting for libelous reflections upon his relations with the demon rum is a Republican organ of small circulation, published once a week in a northern Michigan town of about 12,500 population. To support his civil actions, there have gone with him into that distant quarter a small army of notable divines, family physicians, eminent blood relatives, ex-governors of states, collectors of ports, ex-ambassadors, distinguished sociologists, and real editors. Over the fate of this mighty expedition sent out to crush a country editor the nation sits in anxious interest.*

As Roosevelt moved through the crowd, he was showered with invitations to speak at virtually every local function. To illustrate how sensitive he was to the goings-on, he was invited to a local showing of a play called *The Prince of Pilsen*. Considering the nature of the lawsuit he had initiated, he said he

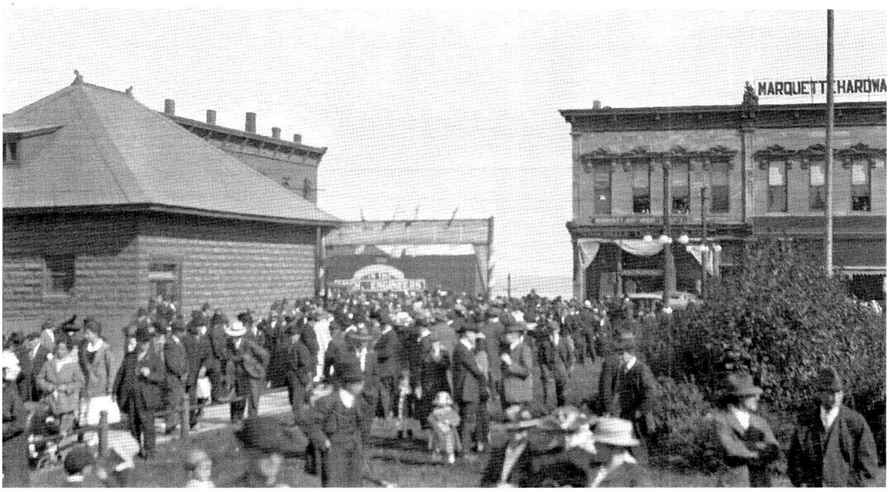

The crowd in downtown Marquette waits for Roosevelt's campaign train to arrive in 1912. Most of the city was present to hear the man who carried a big stick speak. *Courtesy of Jack Deo of Superior View Studio.*

felt compelled to turn that one down: "I don't want anyone confusing Pilsen with Pilsener," a type of beer.

Jay Hayden, the reporter who covered Roosevelt's campaign, was also assigned to cover the trial for the *Detroit News*. He observed as Roosevelt came to Marquette, "I recall the enthusiasm of the waiting crowd for Roosevelt, his affability and kindness, his absolute assurance."

A crowd followed him everywhere. There were twenty reporters assigned to cover the events of the trial from all over the world. Newspapers in London, England, followed the goings-on by running sidebar stories on the makings of American cocktails. Teddy Roosevelt was international news and perceived as a living, breathing American cowboy. It held a certain appeal.

The trial had implications for the press as well. If Roosevelt succeeded with his libel suit, the press would be on notice. It could change how campaigns were reported and candidates were criticized. Supposition, rumor and gossip were under legal assault. No more would journalists be able to sling accusations freewheeling in the newspapers—not simply against Roosevelt, but this trial could set a precedent for all public figures. Roosevelt had put the press on notice everywhere.

While in Marquette, Theodore Roosevelt roomed with his friend and fellow conservationist George Shiras 3[d]. At the time, Shiras had become a world-renowned wildlife photographer. His photographs had been published by *National Geographic*, which at the time was a text-only magazine. Shiras started the magazine's reputation for natural photography with the first appearance of his work in 1905.

Shiras became a regular contributor to *National Geographic*. In the July 1906 issue, some of Shiras's most breathtaking wildlife photos were published. Shiras had developed a method of outdoor photography that consisted of trip wires and bait, which got animals to trip the wires, and they would take their own pictures. The results were astounding, revealing wildlife behavior never before seen.

The issue was so popular that it had to be reprinted to meet the demand. After seeing the July issue, then-president Theodore Roosevelt was so inspired that he penned a letter to Shiras:

> *Dear Mr. Shiras,*
> *I have been looking through your photographs in the* National Geographic Magazine. *Now, my dear sir, no other work you can do (not even going to congress; still less, writing articles for pamphlets or magazines utterly evanescent in character) is as important as for you to write a big book—a*

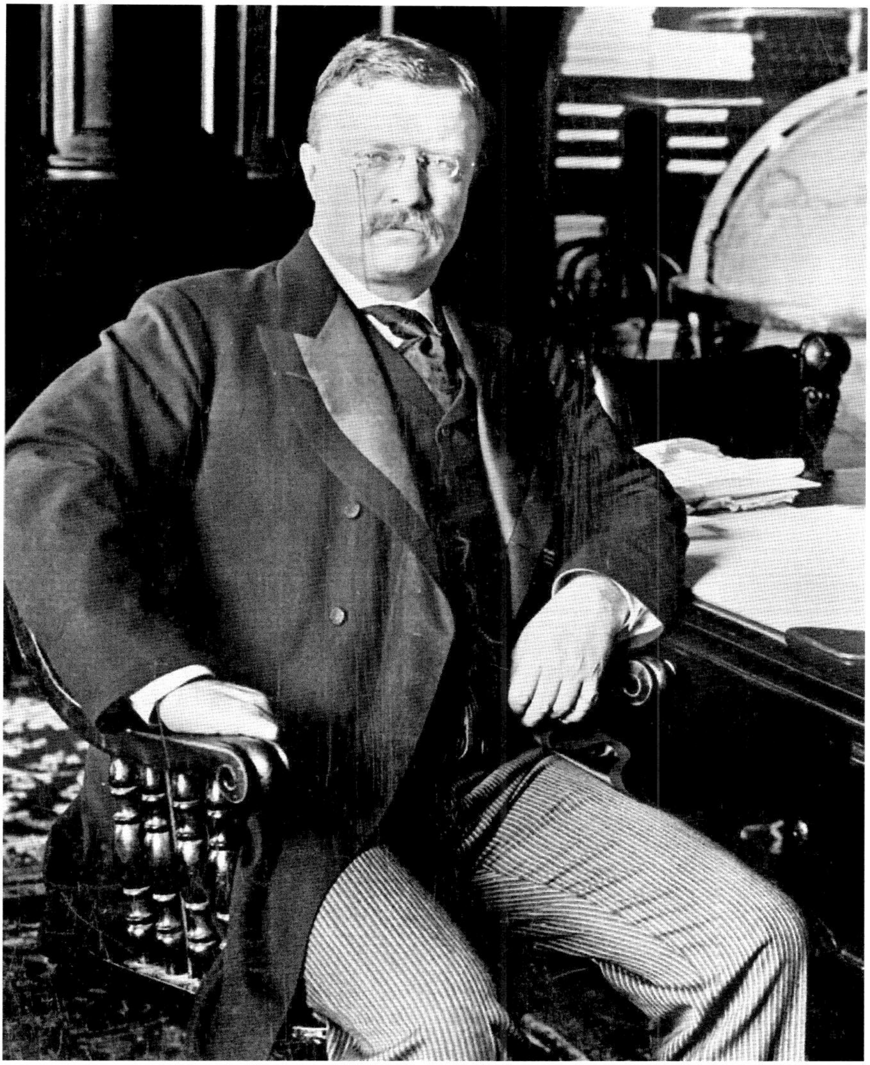

Portrait of a President. This is a publicity photo of Roosevelt. One of these was given to George Shiras 3[d] as a token from Roosevelt. Roosevelt was a big supporter of Shiras's wildlife photography. *Courtesy of Jack Deo of Superior View Studio.*

book of bulk as well as worth, in which you shall embody these pictures and the results of all your invaluable notes upon the habits, not only of game but of the numerous other wild creatures that you have observed. I feel strongly that this country stands much more in need of the work of a great outdoors faunal naturalist than of the work of any number of closet specialists and microscopic tissue cutters. Do go ahead and do this work!

Sincerely yours,
Theodore Roosevelt

Shiras had been a member of Congress and had declined reelection in 1904, deciding to pursue the life of a photographer and naturalist.

Roosevelt's letter was the beginning of a relationship that would last until his death. The two would work together on many projects as conservationists, and Shiras and his camera would accompany Roosevelt to Africa. Shiras was an associate member of the Boone and Crockett Club, which Roosevelt founded.

When Roosevelt came to Marquette in his 1912 campaign, he shouted to Shiras in the crowd, "Did you get that picture of the beaver yet?" "Not yet," was Shiras's reply. Shiras would get the picture, and it would be the very first photograph ever taken of a beaver chewing on a tree.

Shiras and Roosevelt were given honorary doctor of science degrees together from Trinity College in Hartford for distinguished work as field naturalists in 1918. The pair had worked tirelessly to dispel the falsities of wildlife behavior and scientifically establish the true facts of the animal world. What Audubon did for birds, Shiras and Roosevelt did for wildlife.

In the book that Shiras would eventually write, *Hunting Wildlife with Camera and Flashlight*, he says this of Roosevelt:

I have a deep feeling of gratitude to Theodore Roosevelt for his enthusiastic approval of the field work I was doing and his urgent insistence, both while he was in the White House and during the remainder of his life, that my observations be published in full. He repeatedly stated that our literature was deplorably lacking in books on hunting, fishing, and exploration compared with what it might have contained had it been enriched by Americans whose experiences warranted such permanent record.

The book would not be published until after Roosevelt's death, but Shiras felt charged with the task by his good friend and finished it as Roosevelt would have wanted. The book is dedicated to the memory of Theodore Roosevelt.

By the time the trial took place, Roosevelt and Shiras were close friends and both heavily involved in their pursuits of conservation and exploration. It was logical that Roosevelt would stay at the Shiras house. It would be his refuge of quiet during his time in Marquette.

4

Nobody Calls T.R. a Drunkard and Gets Away With It

The first day of the trial was May 26, 1913. The bailiff officially called the case at 2:00 p.m. with Judge Richard C. Flannigan. Earlier in the day, Flannigan had received a telegram from Minneapolis, Minnesota. It said, "Don't let this sensational trial continue until I have arrived." It was signed "Jacob Miles."

Judge Flannigan turned the message over to Frank Tyree, one of Roosevelt's Secret Service men when T.R. was president. He was along to testify as a witness. He contacted the police chief in Minneapolis, who immediately investigated. He wired back that Miles had been arrested and was considered insane. With the assassination attempt less than a year past, no one was taking any chances with Roosevelt's safety.

The first order of business was the selection of a jury. It would not be an easy one. Both sides had strong feelings about who they would accept as a candidate for the jury.

William Belden, Newett's attorney, wanted assurance that Roosevelt's fame wouldn't influence the jurors. Newett sat between his attorneys, straight and stiff. He was sick, and it showed. He was scheduled for surgery as soon as the trial concluded. What he was sick of was never mentioned in the accounts. Whatever it was, it appears, he beat it. He would live another fifteen years after the trial's end.

Conversely, Roosevelt's attorney, James Pound, wanted to be sure that Roosevelt's inflammatory stand on the iron industry didn't prejudice jurors either. With Cleveland Cliffs Iron Company being the major employer of

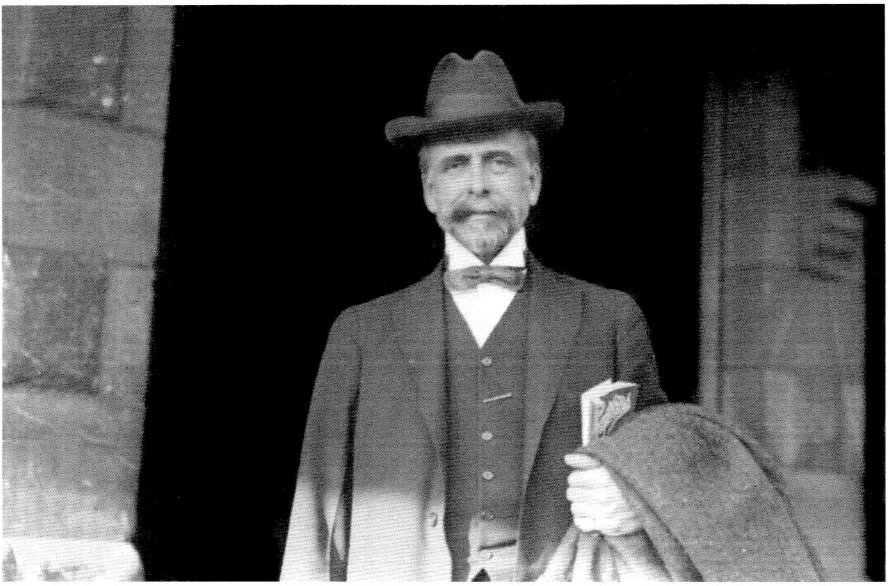

The Honorable Judge Richard C. Flannigan presided over the case of *Roosevelt v. Newett.* This was taken just outside the courthouse during the trial in 1913. This is a rare photo from the scrapbook of George Shiras 3d. *Courtesy of the Marquette Regional History Center.*

the area, there could easily be some local animosity. Newett was a local and had many friends in the community, and the *Iron Ore* newspaper held considerable influence among its readers. It boasted a circulation of three thousand. Attorney Pound wanted to be sure some of Newett's allies didn't infiltrate the process too.

Thirty-six men were brought into the courtroom, twelve of whom would have the opportunity to decide the reputation of a former American president. The importance of what they were about to do had not escaped the citizens of Marquette.

The courtroom filled, and many tried to gain entrance, but they were turned away. The upper galleries of the courtroom were filled with women of the community, while those concerned with the case occupied the seats below. Spectators were allowed to fill the remaining seats. Roosevelt arrived from the Shiras house in an automobile flanked by two Secret Service men.

Judge Flannigan called the court to order, and the bailiff officially called the case of *Roosevelt v. Newett.* The jury selection began.

First to be called was Joseph Robear from Ishpeming. He was a teamster and had been for twenty-five years. Attorney Pound asked the first questions:

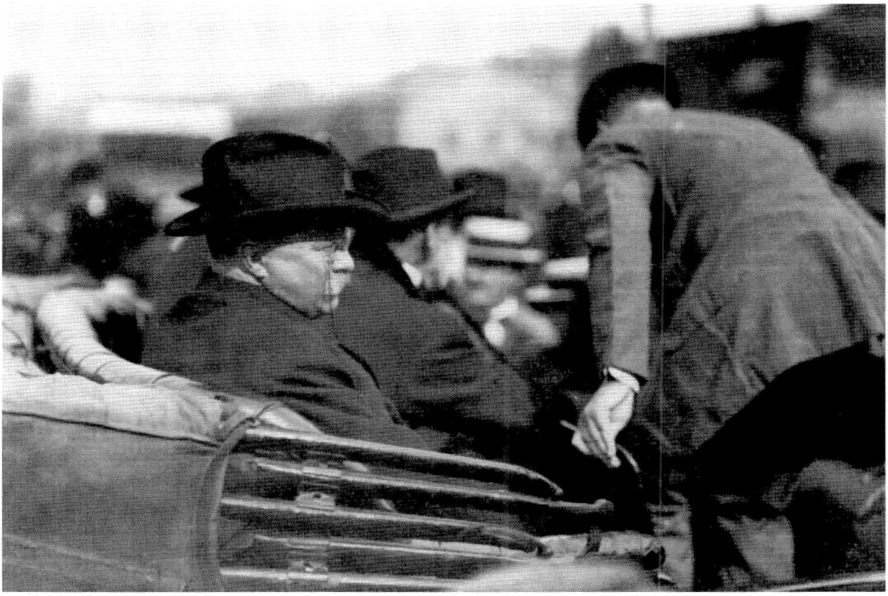

Roosevelt riding back to the Shiras house after a day in court. The man climbing into the vehicle is likely Secret Service. Though Roosevelt allowed himself to be accessible to the public, the assassination attempt the previous year had tightened his personal security. Secret Service agents attached to Roosevelt also testified at the trial as to Roosevelt's sobriety. Their constant presence in Roosevelt's life made their testimony vital. *Courtesy of the Marquette Regional History Center.*

"Does the fact that the libel occurred in Ishpeming influence your opinion on this matter?"

"It does not."

"Do you look upon the publication of such a statement about Colonel Roosevelt as a joke?"

"No, I wouldn't consider it a joke."

Next, William Belden, Newett's attorney, asked Robear, "Would the fact that the plaintiff has been a president of the United States affect you? Would it affect you any different from what you would feel if he was a plain citizen like Mr. Newett?"

"No, that fact wouldn't affect me."

"You wouldn't be influenced by the fact that he has been president?"

"No, sir."

It was agreed Robear was acceptable from both sides. The next couple of interviews didn't go as well.

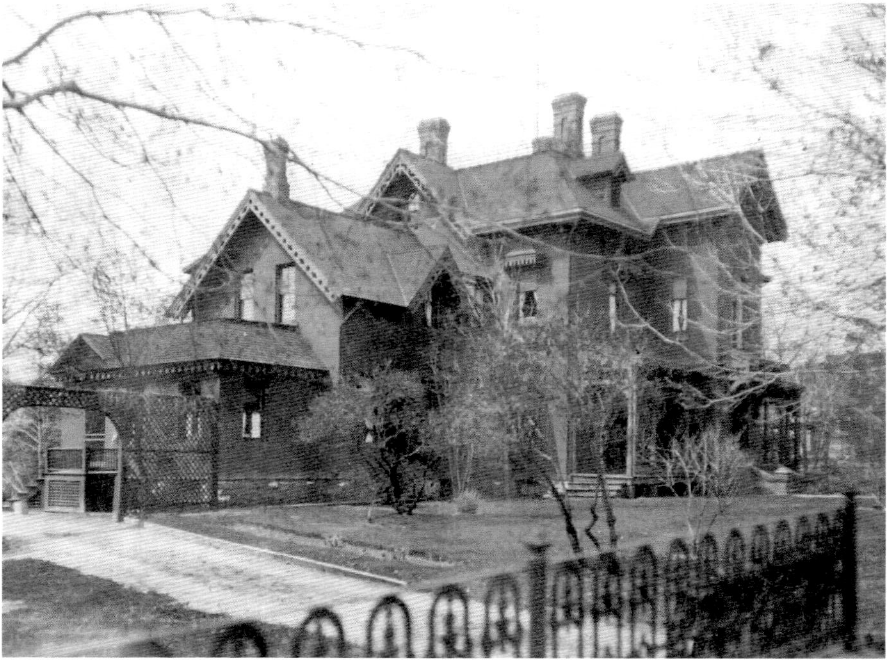

The Peter White house, home of George Shiras 3^{d}, son-in-law of Peter White. Shiras was a close friend of Roosevelt's, and it was the Shiras/White house that Roosevelt stayed at during the libel trial of 1913. Shiras was well known for his innovative techniques for wildlife photography, which was featured prominently in *National Geographic Magazine*. He was the publication's first wildlife photographer. The photo is from Shiras's own scrapbook. *Courtesy of Marquette Regional History Center.*

Hugo Erickson of Ishpeming was next to be examined by the attorneys. He did not know Newett the defendant and had no business or personal relationships with him.

"Would it make any difference to you that one party to the suit was a president from the state of New York and the other a neighbor?" began Attorney Pound.

"No," responded Erickson.

"If a newspaper declares someone is guilty of murder or burglary, would you regard it as a joke or a serious matter?"

"Serious."

Next, Attorney Belden asked, "Do you have a preconceived opinion of the case?"

"No."

"Have you any prejudice as to the right of a newspaper to comment on the actions of a public character?"

"No."

"If we should prove that what the defendant wrote against the plaintiff was true, would you have prejudice against the defendant?"

Erickson was completely stumped by the question. Even rephrasing by Belden didn't help. Erickson just couldn't understand the question. Judge Flannigan suggested that Erickson be dismissed. It was a sign of the long process that was ahead.

The next man on the list was an auto repairman also from Ishpeming. Charles T. Rutledge was the son of a minister.

"Are you a subscriber of Mr. Newett's paper the *Iron Ore*?" Attorney Pound again began the interrogation.

"Yes, my father is; I know Mr. Newett and repair his automobile."

"Your friendship with the defendant wouldn't bias you, in the trial of the case?"

"No, I think I could try it fairly."

"What do you think about the law of libels? Is it a joke or a serious matter?"

"I think it is serious."

"If a man publishes something untruthful about another, you think he should be punished?"

"Yes, I do."

That was the end for Rutledge. He was deemed unacceptable for the jury.

Through the afternoon they ran. Jury prospect after jury prospect was rejected. Judge Flannigan recessed for dinner but reconvened the court for a night session. It ran longer than the afternoon. Man after man was rejected. Several were pressed by the lawyers and admitted they had preconceived opinions that would take evidence to remove.

August Brodin, a prospective juror, came in wearing a Bull Moose pin on his lapel. This got the attention of the two dozen reporters who were intently following the proceedings. They whispered to each other, "Pipe the Bull Moose pin." This caused some quiet snickering in the courtroom. Immediately, Newett's attorney, Belden, started questioning him, not bothering to veil his attitude that he didn't want Brodin serving on the jury: "Have you ever had cause to dislike the defendant?"

"None that I can remember."

"Was there ever anything about you printed in the *Iron Ore*?"

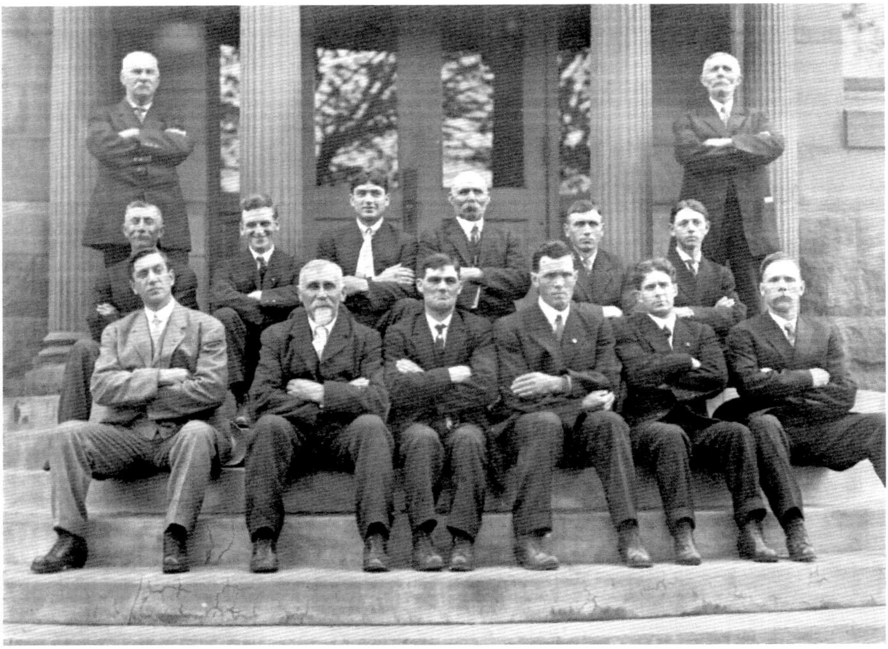

Pictured is the jury at the *Roosevelt v. Newett* trial. We do not know which name to attach to the individual men in the image, so the names are in order of their acceptance into the jury: Joesph Robear, teamster; William Fassbender, farmer; Robert Bruce, woodsman; William Pryor, locomotive fireman; Thomas Howard, farmer; John Johnson, farmer; W.H. Mathews, mining clerk; William Sharpe, teamster; Gus Paulson, blacksmith; Andrew Johnson, farmer; John Fredrickson, miner; William Garrow, miner. *Courtesy of Jack Deo of Superior View Studio.*

"No, nothing."

"You dislike Mr. Newett, do you not?"

"I don't think about him."

Eventually, Belden was able to get Brodin to admit that he did have strong feelings about the case, causing Judge Flannigan to dismiss him. Another of the prospective jurors claimed that he could render a fair decision "regardless of the evidence," which sparked some amusement among those in attendance.

Before the day was over, forty-five candidates had been interviewed to fill out a twelve-man jury. The two head attorneys agreed that they were satisfied with the jury. All of the prospective jurors were asked if they could contribute the one to two weeks the trial might take out of their lives without undue hardship.

Consisting mostly of laborers, the selected jurors consisted of three miners, two teamsters, four farmers, one blacksmith, one locomotive fireman and a lumberman. They were selected as follows: Joseph Robear, Ishpeming, teamster; William Fassbender, Marquette Township, farmer; Robert Bruce, Powell Township, lumberman; William Garrow, Diorite, miner; William Pryor, Marquette, locomotive fireman; Thomas Howard, Chocolay Township, farmer; John A. Johnson, Skandia, farmer; W.H. Mathews, Ishpeming, clerk mining office; William Sharpe, Negaunee, teamster; Gus Paulson, Wells Township, blacksmith; Andrew P. Johnson, Humboldt Township, farmer; and John Fredrickson, Negaunee, miner.

The men were then remanded to the undersheriff; they would not be allowed to leave the building until the end of the trial. Cots were set up for the jurymen in the anterooms of the courthouse. It would be their home for the duration of the case. It would be the first time a jury had ever been housed at the courthouse.

The jury members were to take their meals under the escort of court officials and would not be allowed to see any newspapers. Judge Flannigan instructed the men of the jury to keep their minds open up to the conclusion of the case, not to discuss it even among themselves while the trial was in progress and to immediately report to the court any attempts to communicate with them or in any way try to influence their judgment. He said they would be allowed to write all the letters they wanted, but they would be permitted to receive letters only after they were inspected by court officials to ensure there was no mention or commentary about the trial. Judge Flannigan wanted the jury to stay untainted at all costs.

With the jury chosen and sequestered, the day's process was over. It was late in the evening, and the court was set to reconvene the next day at 9:00 a.m. There was a lot of talk in the city; it was rumored that Colonel Roosevelt would be one of the first witnesses to take the stand.

5

Whose Bully Pulpit Is It Anyway?

The morning came early, and the toll from the late night showed on many of the faces, including those of the attorneys. The jury, being farmers and workingmen, were up and around, used to early hours.

Judge Flannigan restricted the entrance into the courtroom to the principals, the bailiffs, the jury, the court reporters, newsmen and a few officials on the main floor. Witnesses were to be admitted when they were called to testify. Two small galleries, similar to theater boxes, holding fourteen chairs were at the rear. They were occupied by community women.

Roosevelt arrived at the courthouse from the Shiras house accompanied by his legal team. The Secret Service men, though they had come to testify, accompanied T.R. as well. The crackpot telegram of the day before had inspired some extra caution.

Judge Flannigan called the court to order, and the bailiff called the first witness. It was Roosevelt. He was sworn in and took the stand. He sat right next to the jury, so close he could touch them. When Roosevelt began to testify, he had the rapt attention of everyone in the room.

Attorney Pound began the examination of Roosevelt: "What is your name please?"

"Theodore Roosevelt."

"You were born where?"

"I was born in New York City, October 27, 1858."

"Your home during your boyhood was where?"

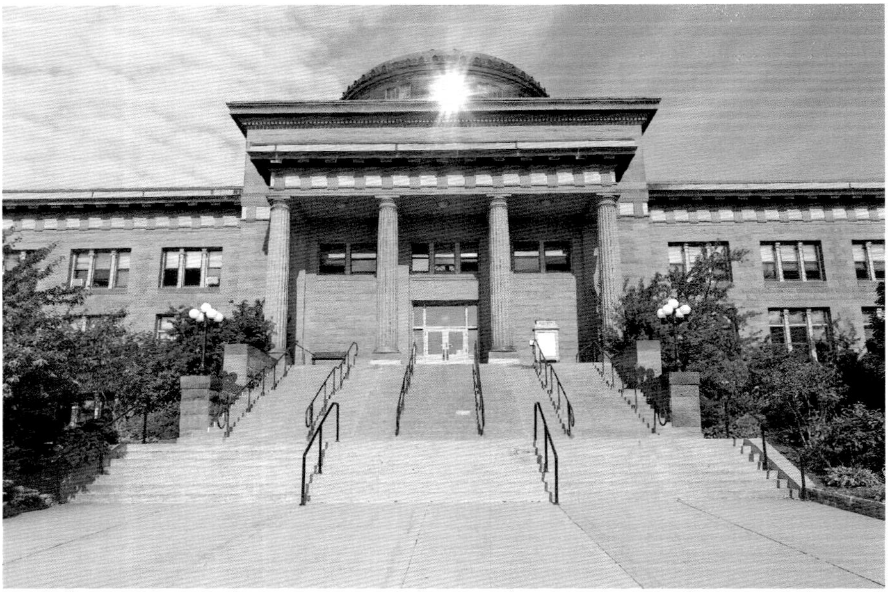

The Marquette County Courthouse as it is today. It has changed very little since the time of the Roosevelt trial and still functions as the hall of justice for the county. *Photo by Mikel B. Classen.*

"New York City in the winter and in the country in the summer, the adjacent country."

"You finished your studies about what year—I mean as a scholar, of course?"

"1880."

"And at that time, did you take a degree?"

"Yes, I took a degree."

"From where?"

"Harvard."

"After graduating from Harvard, what did you devote your attention to?"

"Writing, and I went into politics."

"What was the first connection you had with politics?"

"I was elected to the legislature in the fall of 1881 and reelected in 1882 and in 1883."

"The legislature of what state?"

"Of the state of New York."

"During that time, and as part of your duty, did you become acquainted with a number of government men that were somewhat new to you?"

"I did."

Attorney Pound had been pacing in front of Roosevelt, but now Attorney Horace M. Andrews, one of Newett's other attorneys, stood: "Just a moment. We object to that, as irrelevant, incompetent and immaterial."

Attorney Pound replied, "I am not going to pursue it any further."

As he sat back down, Attorney Andrews said, "We object to it."

Pound looked at the judge, appealing, "I think I am entitled to that much."

Flannigan replied, "I have some question about it, Mr. Pound."

Pound blustered, "Do I understand that in case of an adverse ruling, an exception is given as of course, or do we have to take it?"

Flannigan replied, "You have to take it."

Pound responded, "Please give me an exception."

Flannigan agreed, and Pound continued with his questioning of Roosevelt. "After your service as a member of the legislature, what was the matter that you were associated with, in connection with the government in any way, either municipal, state or national?"

Roosevelt answered, "I was appointed civil service commissioner in the spring of 1889. I served until the spring of 1895. I then was appointed police commissioner. I was president of the police commission of the city of New York until the spring of 1897. I was then appointed assistant secretary of the navy, under President McKinley, and I served until the spring of 1898. I then resigned to accept the position of lieutenant colonel of the volunteer cavalry, commonly called the Rough Riders, in the war with Spain. I served through the fight at Santiago, in which the regiment lost—"

Attorney Andrews again interjected, "I object to that."

Attorney Pound continued unabated, "And after Santiago?"

Roosevelt answered, "I was promoted to colonel."

Pound tried to proceed: "How many times were you under fire?"

Roosevelt began, "I was—"

Again Andrews objected: "I object to that as irrelevant and immaterial."

Judge Flannigan answered Andrews, "I will hear you." He addressed Pound, "Why do you think that is competent?"

Attorney Pound answered, "I think it is a part of the man's history."

Flannigan ruled, looking at the jury: "You may take the answer."

With the objection over, Roosevelt continued again: "I was under all the fire there was in Cuba. I was in the Las Guasimos fight and in the Santiago fight, in which the number of officers lost was over one-third and over one-fifth of the men killed or wounded."

Pound then asked, "You were wounded, Colonel?"

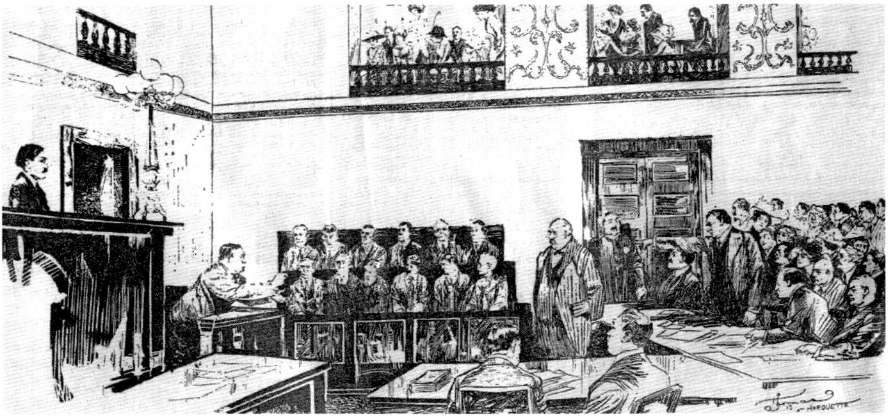

From the *Detroit News*: "Roosevelt on the Stand." This courtroom scene as Colonel Roosevelt testifies was drawn at Marquette by editorial cartoonist Burt Thomas, who was covering the trial with reporter Jay Haden. It appeared on page one of the *News* on Thursday, May 29, 1913. At the extreme left is Judge Flannigan. Colonel Roosevelt is leaning forward in the witness chair, parrying words with Attorney Horace M. Andrews of Cleveland, who is cross-examining him. James H. Pound, counsel for Roosevelt, stands at the corner of the jury box, following the examination closely. At Andrews's right is Attorney William P. Belden, chief defense counsel. On Andrews's left is defendant George A. Newett's other attorney, Thomas Clancey, and beside Clancey, at the end of the table, is Newett himself. At the Roosevelt counsel table, nearest the judge's bench, sits Attorney W.S. Hill of Marquette, and with him is Attorney William H. Van Benschoten of New York. The galleries above the courtroom are filled with women of Marquette, Negaunee and Ishpeming, who made a social affair of the trial. *Courtesy of the Marquette Regional History Center.*

Again, Andrews interrupted: "I object to this statement." But it was too late; Roosevelt had pulled up his shirt cuff to show a very visible scar on his forearm.

Pound ended it and began again, "That will be sufficient. At the expiration of the contest in Cuba, where did you go? Were you mustered out?"

Roosevelt answered, "The regiment was mustered out in September 1898, at Montauk Point, and I was elected governor of New York that same fall."

"How long did you serve in that office?"

"I served for two years, and while governor, in the spring of 1900, I was nominated as vice president with President McKinley and served as vice president until in September 1901, when President McKinley died because of the bullet wound inflicted by Czolgosz. I then became president, and I served the three and a half years of the unexpired term for which McKinley was elected. I was nominated and elected president in 1904 and served until March 4, 1909."

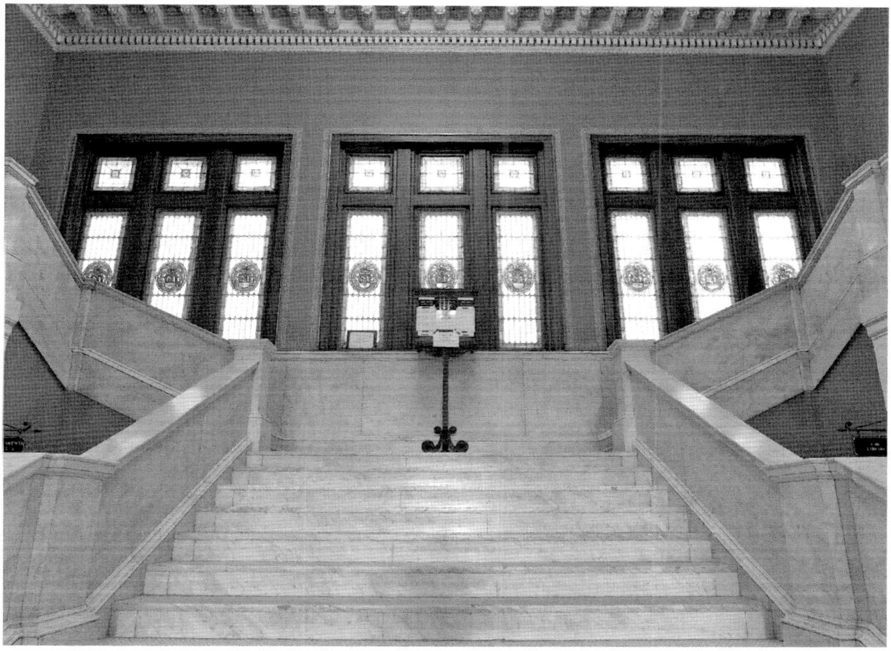

The stairway leading to the Marquette County Courtroom. This is how the stairs looked when Roosevelt climbed them in 1913. *Photo by Mikel B. Classen.*

"Now, Mr. Roosevelt, after the expiration of your term of office as president, what did you then next devote your attention to?"

"Two weeks and a half after I left the White House, I went to Africa, as the head of the Smithsonian expedition for the National Museum, an expedition to collect natural history material for the National Museum at Washington. I was a year on that expedition. I was eleven months in Africa, entering at Mombasa on the east coast, and going into the middle of Africa and down the Nile. I then spent about two months in Europe and came home, reaching home in June 1910. Since then, I have been connected with the *Outlook*."

"In that connection with the *Outlook* [a magazine Roosevelt worked closely with], have you had occasion to travel somewhat over the United States since your return?"

"I was independent of that connection."

"You have though?"

"I have traveled back and forth across the United States since I returned."

"Now, did you ever live in the West?"

"I did."

"Tell me when."

"From 1883 to 1896."

"What portion of the West were you in?"

"I lived on a cow ranch in the short-grass country of the Little Missouri."

"In what state?"

"In what was the Territory of Dakota when I went there but now is the state of North Dakota. I lived there at intervals for two or three years during the major part of the time, but for the latter part of that time, I was only out for a few months each year."

"Now, during the year 1912, did matters so shape themselves that you participated in the campaign of 1912?"

"I participated in it?"

"In the month of October 1912, were you in the state of Wisconsin?"

"Yes."

"I call your attention to the date October 11, 1912. Do you remember that date?"

"If you could identify it by some incident; I don't remember that date."

"I ask you to look at this paper and look at the date, if you please?"

"Yes."

"Did you ever see a copy of that?"

"I did."

"With reference to October 11, this date, as near as you can tell, when was it that you saw that?"

"It was within a few days, but I cannot remember whether it was before or after I was shot."

Once again, Attorney Andrews interrupted, "I move to strike out the answer."

The judge replied, "The answer may stand."

Attorney Andrews was still persistent: "I except. I move to strike out the latter part of the answer, relating to when he was shot as incompetent, irrelevant and immaterial."

Pound argued, "I only offer it as fixing the date."

The judge let it stand: "You have an exception, Mr. Andrews. Let the answer stand."

Pound went back to his line of questions: "Where were you when your attention was first called to this article?"

"That is what I'm not sure of. My memory is that I was in Mercy Hospital at the time, but it may have been before I had gone there. That is the only reason I mentioned about being shot, that I don't remember."

"Mercy Hospital is where?"

Again, Andrews objected: "I object to that as immaterial in this case."

Judge Flannigan overruled the objection.

"Note an exception," replied Andrews.

"What is the answer?" Attorney Pound tried to get back on track.

"Mercy Hospital is in Chicago," answered Roosevelt.

"You are in doubt as to the other place, where you say you were shot. Where were you shot? What city?"

Again, Newett's attorney interrupted, frustrated. They didn't want the jury leaning toward Roosevelt just out of sympathy for the assassination attempt: "I object to that! I don't think that fixes any date, and the evident purpose is, it seems to me, to put in here something entirely foreign to justice, in this case."

Attorney Pound shot back, "It is strictly competent. He identifies it in connection with one or the other of two places."

Judge Flannigan addressed Pound, "What do you expect to claim for this testimony?"

"Nothing more than to fix the division of time beyond peradventure."

Flannigan let Pound continue and addressed the jury: "For that one purpose you may take the answer."

Andrews remarked, "Note an exception."

Roosevelt answered the question: "Milwaukee."

Andrews again interrupted, "That has no relevancy to this case."

The judge was getting tired of Andrews's continuing objections: "We have ruled on that proposition, and you have an exception."

Pound had walked over to the plaintiff table and picked up a newspaper, a copy of the *Iron Ore*. He handed it to Roosevelt and went back to questioning him: "Calling your attention to this article, did you read it?"

"I did."

"How did you feel about it?"

"I felt indignant."

"Were you at that time able to recall the gentleman whose name is at the head of this sheet as the owner?"

"I think not?"

Pound turned to the defendant's table and addressed Newett's lawyers: "You gentlemen do not raise any question about this paper, do you?"

Attorney Belden shook his head and replied, "No."

Pound continued, holding up the copy of the paper that held the now-infamous editorial: "I think we may as well here as any place introduce this, and I will offer this article in evidence."

Belden answered, "There is no objection to that."

Attorney Pound continued, "I will read that portion of it which I deem to be pertinent, from the article under the caption 'The Roosevelt Way.' I will offer in evidence the head of the editorial page down to and under the flag."

Judge Flannigan granted him permission: "You may read to the jury at this time just what you offer."

Attorney Pound began: "For identification, I will read: '*Iron Ore*. Two dollars per year in advance. George A Newett, Publisher. Entered at the post office at Ishpeming, Michigan, as second class matter. Sunday, October 12, 1912.' Now I read as to the libel."

Pound proceeded to read the entire editorial, "The Roosevelt Way," to an attentive jury. When he finished, he turned again to Roosevelt: "You are now fifty-four years of age, as I figure it out?"

"I am."

"I ask you, since your arrival at the age of manhood, what is the fact as to whether you have ever been under the influence of intoxicating liquors of drugs?"

Roosevelt leaned forward in his chair. They had come to the crux of things. His answer was calculated: "I have never been drunk or in the slightest degree under the influence of liquor."

Pound went on, giving the colonel the opportunity he wanted: "Now, I wish you to describe in your own way to the jury, what, if any, use you make of liquors, spirituous or malt, since your manhood, in your recollection."

"I do not drink either whiskey or brandy, except as I shall hereafter say, except as I drink it under the direction of a doctor. I do not drink beer. I sometimes drink light wine."

"Let me ask you right here, have you ever indulged in porter on any occasion?"

"I never drank liquor or porter or anything of that kind. I have never drunk a highball or cocktail in my life. I have sometimes drunk mint juleps in the White House. There is a bed of mint there, and I may have drunk half a dozen mint juleps a year, and certainly no more."

"All at one time, or was it just a…some people here may be in the courtroom for the first time, and I desire you to state the facts."

"I never drank but one mint julep at a time. I doubt if I have drunk a half dozen a year. I doubt if I drank a dozen during the entire seven and a half years there. Since I left the presidency, in the four years since I left, I remember I have drunk two. One at the country club in St. Louis, where I simply touched a mouthful, and one at Little Rock, Arkansas, where there was a dinner given by the governor of the state and others, and they passed a loving cup of mint juleps, and I drank from that as it passed."

"What is the fact, I may be too anxious about it, but I want to know whether there was ever a time that you drank two mint juleps inside one day of twenty-four hours. Would that much time pass between the half dozen you have tasted in the four years you were drinking in the White House?"

"I should say, Mr. Pound, it would be perfectly possible for me to take that much on some occasion in the past thirty-four years or nearly, to drink two mint juleps together, but I don't think so. I think not."

"Now, you say you drink light wine. Will you describe the amount?"

"I don't like beer, and sometimes I will go to a friend's house where they will ask me to drink beer or whiskey, but instead of drinking beer or whiskey, I will drink light wine, a glass or two glasses. At home, at dinner, I may partake of a glass or two glasses of white wine. At a public dinner, or a big dinner, if they have champagne, I will take a glass or two glasses of champagne, but I take it publicly just as much privately."

"While you were president of the United States, will you tell us what the fact was as to whether there were a number of dinners, three or four annually known as state dinners?"

"Yes."

"Or diplomatic dinners?"

"As you understand, there are in the White House every year three or four dinners, sometimes more, which are called state dinners, regular functions to which members of Congress, foreign ambassadors and others are invited. I also dined, as every president does once a year, with each member of his cabinet, and at those dinners, as far as I can remember, without exception there was champagne. Outside of that, in the White House, I have drunk light wine, and not usually at all if we were alone. If there were guests, I might drink a glass or two of light wine, and I might not drink anything, and in the White House, I never touched brandy or whiskey except as I have described it in connection with the mint juleps or under the doctor's direction."

"Just one question right there. Has there been, since your experience in Cuba, anything that occurs to you occasionally in the way of illness or chills?"

"Oh, I have had, ever since the Cuban campaign, I have now and then had a slight attack of fever, and which will be brought on usually by overexertion. I had a couple of attacks of fever in Africa. I had one attack when in the Rocky Mountains. In the Rocky Mountains when I had an attack of fever, it lasted two or three days, and then it came on again, and in the first attack of it, the doctor gave me a drink of whiskey, but I told him I didn't think it did me any good, and when the relapse came, I asked him to not give it to me, and I took hot tea instead. In Africa, the expedition took with it a case of champagne, a case of whiskey and a bottle of brandy. I never touched one drop of either the champagne or the whiskey. The champagne was used purely for certain members of the party who got dysentery and fever towards the end of the trip, and it was also used for certain elephant hunters and traders and missionaries whom we met who were sick with fever or dysentery, and it was only used for such purposes. The whiskey was used chiefly for such purposes, but some members of the party drank it otherwise. I never touched a drop. The brandy bottle, I was the only person that drank from the brandy bottle."

Attorney Andrews objected, "I think this is not responsive to any questions asked."

Pound turned toward him: "Yes, I asked him what the experience was in Africa."

Judge Flannigan wasn't inclined to stop Roosevelt and his tale of African exploration: "You may proceed." Looking at Andrews, he said, "You may have an exception if you care for it."

Andrews nodded, "I except to the ruling."

Roosevelt continued with his testimony: "The only brandy I drank was at the time of my two fever attacks, by direction of the doctor, about seven tablespoonfuls. I think most of it was in the first fever attack, and then I told the doctor, Dr. Mearns, when the second fever attack came on—"

Andrews interrupted, "The conversation with the doctor I ask to have excluded."

Attorney Pound acquiesced, "That probably would not be proper."

Roosevelt picked up where he left off: "After the second fever attack, I refused to take any more and took tea, and during the eleven months that I was in Africa, the only liquor, brandy or whiskey that I took were those seven tablespoonfuls of brandy."

"How many men were in your party that started with you from Mombasa?"

"Five."

"Was it added to by guides or others as you traversed the country?"

Andrews objected, "That is objected to as immaterial."

Flannigan had developed a pat response: "You may have an exception."

Roosevelt had the courtroom's complete attention: "Two other men, two professional elephant hunters, joined us."

"Where was it that you left Africa proper before you reached Egypt?"

"Khartum."

"You went in at Mombasa and came out of Africa at Khartum?"

"Yes."

"From Khartum, your progress down to the Mediterranean was by what method?"

"It was by railroad. At Khartum I was met by my wife and daughter and son and various friends and certain newspapermen, who continued with me until I came home."

"In the basket of champagne, how many bottles would be included?"

"You mean the case of champagne?"

"Yes."

"I don't know. I never saw it."

"What I am getting at, was it six or eight or ten or twelve bottles?"

"I couldn't tell you. I don't know how many bottles there were."

"I understand you that you do not claim to be a prohibitionist or total abstainer?"

"I am not."

"How would you describe yourself as a user of any kind of liquor, free, or—"

"I use—"

"Just a moment," Andrews protested.

Pound continued anyway: "How would you describe yourself as a user of liquor, either spirituous or malt, as to being a free consumer of them to excess or abstemiously?"

"That is objected to because it is leading and incompetent," Andrews again protested.

The judge agreed but let Pound continue: "That seems to be clearly incompetent. I will give you an exception."

Roosevelt spoke up, "Will you allow me to make my statement as to how much I use?"

Pound replied, "Go ahead and state it."

"I can begin by saying that I am not a total abstainer, but that I am very abstentious, very abstentious in the use of wine at all, that I know of no man in any profession—" Roosevelt was interrupted.

"Just a moment." It was Andrews.

Pound replied, "Well, that may not be material."

Andrews took the opening: "I ask that it be excluded and ask that the jury be instructed not to consider it."

Judge Flannigan turned to the jury: "The last answer will be stricken out, the last part of it. Now, gentlemen of the jury, as this suit progresses, if a question is answered and the court strikes out that answer, you will disregard the answer. It will drop right out of your minds."

Attorney Pound turned back to Roosevelt: "Now you may just describe your consumption of liquor in the last fifteen years."

Roosevelt attempted to begin again: "When I went to Cuba—"

Andrews again halted the testimony: "May I suggest, if the court please, I don't want to interrupt Colonel Roosevelt unnecessarily, but it seems to me this matter has all been gone over, and while it may be entirely competent, I doubt the advisability of repeating it."

Obviously annoyed, Judge Flannigan responded, "Repetition becomes necessary on account of interruptions. You may go on." The last was addressed to Pound.

"Just describe it in your own way," Pound said to Roosevelt.

"When on the campaign in Cuba, I drank nothing. I had no whiskey or brandy with me. Before I left, several people sent me whiskey and brandy bottles, and I asked the colonel of the regiment, Colonel Wood, what he would do with it, and they were turned over to the hospital. Since coming back, I have told you how much I drank in the White House. That same statement applies when I have been home and applies while I was governor. May I describe the routine of the White House?"

"As briefly as you can. I think we would all like to know that," said Pound, knowing that Roosevelt had an attentive audience.

"I would get up in the morning at about eight, breakfast with Mrs. Roosevelt and the children. Walk around twenty minutes, walking around the White House grounds, and from there, go down to where my private secretary, Mr. Loeb, was and take up—"

"The offices were where in reference to the capital?"

"The office is connected with the White House, about two hundred feet away. I would take up my work with Mr. Loeb and answer the mail, and senators and congressmen would come in. I would have a cabinet meeting. Every moment would be occupied up to about one o'clock. At one o'clock, unless I had shaved myself, I would be shaved, in a little room between the room where I sat and the room in which Mr. Loeb sat. I would quite frequently, instead of shaving myself, get somebody to shave me, because I could do work while I was being shaved. I had to economize every spare moment. Mr. Loeb, or sometimes a senator or congressman or someone else, would come in and see me while I was being shaved. Usually Mr. Loeb was in at the same time, and I would dictate letters to him, and then as soon as I was finished shaving, I would go and wash my hands, Mr. Loeb usually

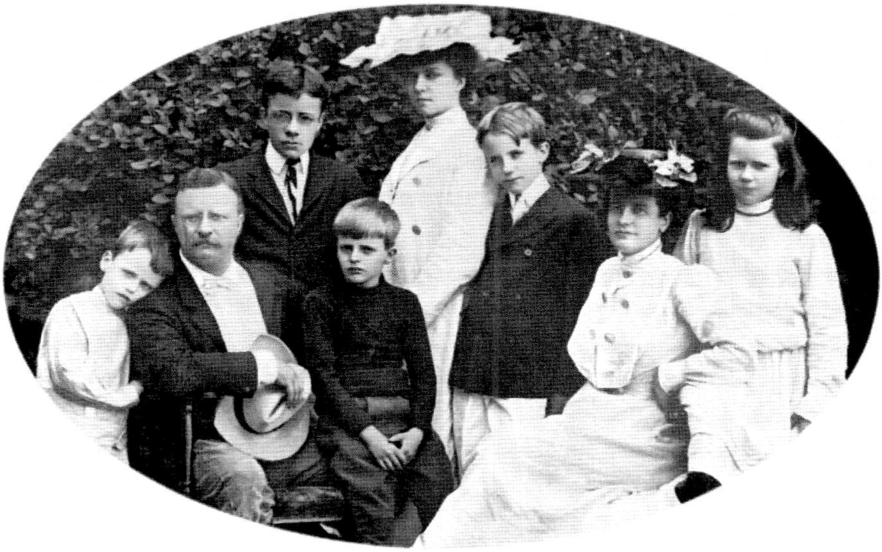

The Roosevelt family (left to right): Quentin (1897–1918), Theodore (1858–1919), Theodore Jr. (1887–1944), Archibald (1894–1979), Alice Lee (1884–1980), Kermit (1889–1943), Edith Kermit (1861–1948) and Ethel Carow (1891–1977). *Courtesy of Jack Deo of Superior View Studio.*

coming in with me and I giving dictation or answering questions as I was finishing washing my hands. Never on any occasion on any day, in the entire time that I was in the White House, had I ever touched a drop of anything prior to lunch, never under any circumstances."

"Then what followed?"

"I would go over and lunch with Mrs. Roosevelt, and sometimes with guests."

"Did the lunch immediately follow your ablutions?"

"As soon as I washed my hands, I would walk right across the terrace and take lunch with Mrs. Roosevelt. Often there would be people there, and almost always, because I had some business to do with them. If Mrs. Roosevelt and I were alone, we had only tea or milk or water, whatever it was, at lunch. If there were any guests, there would usually be light wine on the table. Immediately afterwards I would again go back to be with Mr. Loeb and finish up the work, and I would go over with any head of a department or anyone who had anything to put before me in connection with the government service. At four in the afternoon, or as near that as possible, I would go out for a ride or for a walk or play tennis and then stop and get back about six. I would then take up the mail that had accumulated in the afternoon and answer that and do any other things that there were to do. I would then spend half an hour with the children and bathe and dress for dinner. If we dined alone, we had no wine and no liquor of any kind on our table. If there were guests and it was a small affair, we would have light wine, of which I have spoken. If it was a state dinner, there would be champagne. If we were alone, as soon as dinner was over, I would again go back to the office, or get Mr. Loeb or some head of a department or any other man with whom I had business up to my library and go over the business with him. And after a cabinet dinner or state dinner, as often happened, I was pressed with work. I would work perhaps half an hour or three quarters of an hour after coming back to the White House. I would then usually sit down and read some book entirely disconnected with my work, so as to get the work out of my head before I went to bed. I would go to bed at about twelve. That was the usual time I was able to go to bed, and at that time not infrequently, Surgeon General Rixey, who was my doctor during those seven and a half years, would be in to see me, when I would go to bed and get up and breakfast as I have described. On these trips through the country—"

Attorney Pound stopped Roosevelt: "Just a moment. I want to finish up with the White House now. During that seven and a half years when you were in Washington, from the time you arose at eight o'clock in the morning until you went to bed at twelve o'clock at night, were you constantly engaged in business excepting that time that you took for recreation?"

"Excepting the time I mentioned for recreation, and excepting the half hour I mentioned when I was with my own family, I was engaged every minute of the day in business, and not excepting those times I was always in sight and in touch with either the members of my own family for the half hour or thereabouts that I have spoken about, or Mr. Loeb and the other people who were associated with me, and that included the two hours' walk that I took. For instance, the Secret Service men were always in attendance, and whenever I went out for any reason, they would go out with me. When I went to church, one of them would follow some little distance behind, and when I went out in the evening on any occasion, they would follow with me. When, for instance, Speaker Cannon had a seventieth birthday celebration in the evening, on that occasion, Mr. Loeb went with me. We keep a record in the White House of the particular hours at which I left the White House and came back to it. When I went to Speaker Cannon's reception, the card shows that Mr. Loeb and I started at about 9:20 with Jim Sloan, the Secret Service man. We went there to the reception and came back about eleven o'clock. At that particular reception, as he was from Jim Sloan's district, Mr. Sloan, the Secret Service man, accompanied me out to the carriage and asked if he could stay, as he had friends there. Mr. Loeb said yes, that he was going home with me, and he went home with me. And at that reception, as at the cabinet dinners, as at every other occasion, I never touched a drop of whiskey or brandy. At that reception, I drank a glass of champagne and I ate a sandwich, and I drank the glass of champagne in proposing Mr. Cannon's health on his seventieth birthday."

"I think the jury would like to know to whom you refer as Secret Service men."

"They are men furnished by the United States government for the protection of the president. Secret Service men are two men, or several men, allotted to follow the president around and prevent assassins and cranks from getting at me. For instance, I would not allow them at one time to go to church with me."

Once again, attorney Andrews objected to the questioning: "Just a moment, I don't think that is competent."

Attorney Pound took his point: "I don't care to go any further than that." He turned back to Roosevelt: "You said you enlisted in the army as lieutenant colonel of the First Cavalry."

"Yes."

"What was your rank when you were mustered out?"

Again Attorney Andrews interjected, "I object to that as immaterial."

Pound explained his position: "I insist that it is entirely competent. Of course, the only object I have to show, if there was any promotion bearing on the impairment of the man's health and bearing on the man's life."

Flannigan agreed: "I think the position in life of the plaintiff is competent and material here."

Roosevelt continued, "I was made colonel. I was recommended for brevet rank and for a medal of honor, and I came back acting as brigade commander, in command of the brigade."

Pound asked, "You say brevet…"

"I was recommended as brevet brigadier general."

"Now, you began to state, and I diverted you a moment ago, in regard to certain trips that you had made. Now if you will take that up in point of order during the time of your presidency, or anterior to that, and describe what there was in the shape of drinking, I will be much obliged to you."

"First, occasionally, but still more as president, I traveled over the country, following exactly the precedents of Mr. Cleveland and Mr.

McKinley and my predecessors in the presidency. On those trips I never, while on the cars, touched any liquor or any wine, except just before I went to bed at night. If it was a short trip, I touched nothing whatever, but if it was a long trip and the strain especially on my throat was very severe, after undressing at night, on the advice of Surgeon General Rixey and Dr. Holbrook Curtis, I would take either one or two goblets of milk with a teaspoonful of brandy in each. I would usually take that and then read for twenty minutes in bed by the electric light, and then I would sleep perfectly, and be all fresh in the morning. And on these trips on the average day, which meant nine days out of ten, I would touch nothing during the day of any kind. For instance, that was the case passing through Michigan last fall. It was the case in Ohio last year. I would touch nothing of any kind until I went to bed. On occasions on that trip, I would stop at some friend's house. For instance, in Toledo, Ohio, after my speech, I went to the house of Mr. Shepley to meet Mayor Whitlock. Mayor and Mrs. Whitlock were there, and Mr. and Mrs. Shepley and Regis Post and myself. On that occasion, Mr. Shepley asked Mayor Whitlock and Mrs. Whitlock and all to take a little supper, and there was champagne. I drank a glass and then found that they had a pitcher of milk, and then I took milk and doughnuts instead."

"That suited your appetite better?"

"That suited my appetite better, and I should say on an average, on those trips there would be about once a week when there would be some experience like that."

"When you had a hard day or a long trip, does that mean the distance, or how about the speeches you were called upon to make, or both?"

"It means both the distance and the speeches. It is very difficult for anyone, who has not been through it, to realize the strain that there is in such a trip. For instance, I remember in Kansas, in two days, I made thirty-seven speeches. I began each day with a speech before breakfast, and on each day, I finished my last speech at eleven in the evening, making nineteen one day, eighteen the preceding day, and often the speeches were in railroad yards. We were running races with railroad trains going by and whistles blowing. Often speeches out in the open air, where there was no chance to save my voice, but I had to try to reach the people on the outside of the crowd who had come from a long distance and who I felt were entitled to hear me if they could, and it was as exhausting an experience as could possibly be, generally, and every man who speaks like that must take care of himself all that is possible."

"Did you feel it was a duty you owed to the people that you should exert yourself to make yourself heard and understood?"

Andrews protested, "I object to that as leading, immaterial and irrelevant."

Attorney Pound rephrased the question slightly: "Why was it that you exerted yourself to the degree that you did?"

"There was no point in my going unless I used my voice to make myself heard."

"Were you two or three times in the legislature?"

"Three times."

"Did you have any experience in New York in relation to the mayoralty?"

"I ran for mayor in 1886 and was defeated by Abram S. Hewitt and Henry George."

"Another question on the subject of these state dinners. At these state dinners, state whether your consumption of champagne did or did not exceed, at the utmost, two glasses."

"It would be either one or two glasses."

"During the last fourteen years, for medicinal purposes or otherwise, how many drinks of whiskey have you had, to your best recollection?"

"Except as I have testified, in connection with the teaspoonful of brandy in the milk at night, and excepting as I have testified to in connection with the mint juleps, and on the advice of doctors, I believe in the last fifteen years, I have not had a dozen. I have not taken whiskey a dozen times."

"I ask you, even on these trips, how many times have you in the last fifteen years carried even a small flask of whiskey with you?"

"For the last fifteen years I have never carried a whiskey flask or a brandy flask with me. In the days on the ranch, we never had whiskey on the ranch, but I used to then have a pocket flask of brandy which I would keep for accidents, but I never used it, and have never used it, and I abandoned carrying it, and I have not carried it for twenty years."

"When you were on the ranch, did you frequent saloons or have any occasion to?"

"I never went into a saloon. I don't believe that I ever went into a saloon in the western country except where it was at a little hotel, where the only two rooms would be a kitchen and a dining room where you had the dining table and everything else."

"And bar too?"

"In the room with the bar, and I don't ever remember of drinking at a bar. Certainly not for fifteen or twenty years, and I do not remember of more

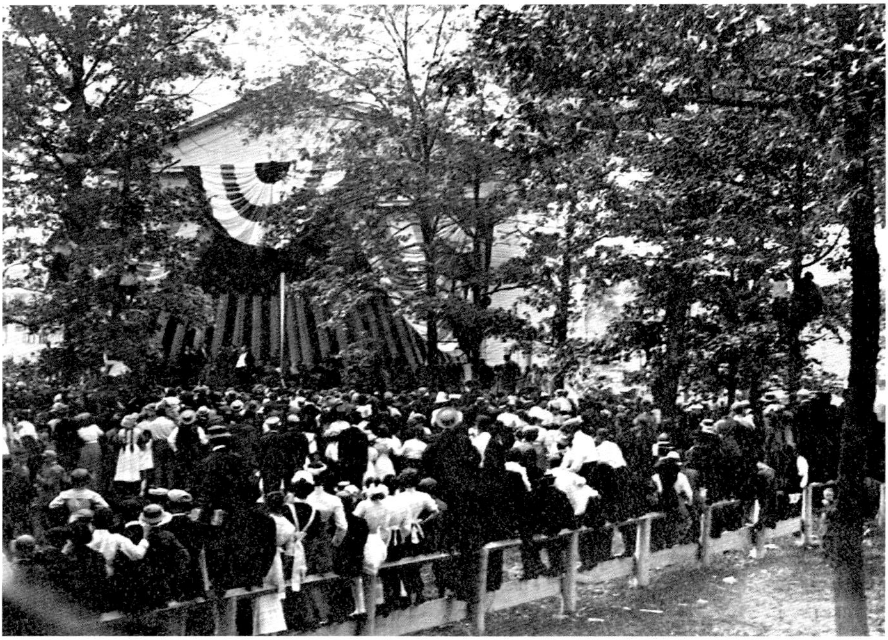

Roosevelt campaigned throughout the Upper Peninsula in 1912. Crowds turned out everywhere in large numbers to see him. His message of anti-corporatism rang strong with the common man. *Courtesy of Jack Deo of Superior View Studio.*

than once or twice going into a saloon, and as far as I can remember, not at all in the last fifteen or twenty years."

"You said you were in the city of Milwaukee—"

"I had no whiskey or brandy in my possession at that time or on any of these trips."

"You said you were in the city of Milwaukee. Did you get out of there without swallowing any lager beer?"

"On this occasion, yes."

"I think you covered it, but I want to be sure. Just one question, whether what you have said of the fifteen years, does that also apply to your methods during the year 1912 in the campaign as well as the rest of the time?"

"It applies exactly! What I have said about the trips while I was president applies exactly to the trips since I have been president including 1912."

"Now I call your attention to the state of Ohio. What is the fact as to whether you made a personal canvass, as strong as you were able, for that state?"

"I did, for nine days. I was nine days in the state."

"Without any surplus words, tell us how you traversed the state, from what point you entered and to what point you went, whether V-shaped or parallel, or how."

"We went around the state, and I think with certain zigzags. We went across the north of the state to Toledo, through the middle and eastern parts, and then the western part of the state to Columbus, and then to Cleveland, and I then spent Sunday with Mr. Garfield at his home and continued the trip. I think on Monday was the last day. It may have been Monday and Tuesday."

"Did you make an address to the constitutional convention in the state of Ohio?"

"I made an address to the constitutional convention."

"About how long before or after the campaign trip for the delegation was that?"

"That address was made in February. I went to Columbus. I had prepared the address five or six days in advance. I went to Columbus and stayed at the house of Dr. Washington Gladden. I made the address and came home. I should say it was a month later that I made my speech in Carnegie Hall, which was practically the opening of the primary campaign, Carnegie Hall, New York."

"There was a time in your life that you thought you had wasted it on the law business, I believe?"

"I studied law for a year after leaving college."

"What I want to ask you is this, as a part of your education curriculum, were you a member of debating societies, or anything of that kind, so as to get a special training as to public speaking, or have you had to rely upon your own experience?"

"No, I never got any training in public speaking until I ran for the legislature."

"Did your first experience come as a candidate?"

"I think my first experience did not come as a candidate. I think it came when I was in the legislature."

"You mentioned Dr. Rixey. Will you kindly tell me a little in detail who Dr. Rixey is and was and how he came to be connected with you?"

"The head of the medical department of the navy is called the surgeon general of the navy, just as the head of the medical department of the army is called the surgeon general of the army. The president usually has either the surgeon general of the navy or the surgeon general of the army as his attendant, sometimes some army or navy doctor other than the surgeon general. When I became president, Dr. Rixey was surgeon general of the navy under President McKinley and was President McKinley's family doctor. He had attended President McKinley through his shooting when he died, and he was transferred to me and became my family doctor all during the seven and a half years that I was in the White House. There was an occasional time when he would be absent from Washington on official business. I think I sent him once for three months to the Philippines on official business, but when he stayed in Washington, he was with me on most of my trips. I should say that he must have been with me nine-tenths of the time that I was president, and he used then to see me every day and often many times a day, and he looked after me to keep me in—"

"He would see you at what times?"

"Oh, it would depend. If he thought that I needed attention, he would see me before breakfast and wait to see me after dinner."

"How late at night has he been at the White House?"

"Twelve or one o'clock. He would wait sometimes and come up after dinner if there was something. You see, the president has got to go on with his business. He can't be sick. He has got to go on with his business, and if I had any trifling ailment, he would have to be with me. If I had an attack of indigestion or anything of that kind, he would have to see that I could go right on exactly as if I didn't have it."

"A sort of physical insurer of the president?"

"Exactly."

"Your family doctor at home in New York, at that time and for many years, has been who?"

"Dr. Alexander Lambert. He has been a close personal friend and the family doctor and has been on trips with me."

Attorney Pound was through. That was the end of his questioning. Judge Flannigan signaled to the defense team: "You may cross-examine the witness."

Attorney Andrews rose and walked to the witness stand. He would do the questioning. He began, "Colonel Roosevelt, you have told us your age and that your boyhood was mainly spent in New York?"

"In New York, my boyhood was mainly spent in New York."

"That is you spent the wintertime perhaps in the city and summertime at your country place?"

"At my father's, where he had a country place."

"By that, you mean a place outside of the city where the family spent the summer?"

"It was generally a farm where the family spent the summer."

"You have never lived in the state of Michigan, have you?"

"I have never lived in the state of Michigan."

"And have never in the last eight or ten years been here, except upon a flying trip?"

"Excepting on a comparatively short trip, I was several days here when I was inspecting the naval militia."

"That was how many years ago?"

"I think it was fifteen years ago."

"So that I would be correct in saying that within the last ten years, you have scarcely been within the state excepting upon a short flying trip?"

"Excepting upon short visits through the state, usually to speak."

"Your acquaintance in Marquette County is somewhat limited, is it?"

"It is limited."

"Do you know any of the residents here?"

"I know Mr. Shiras."

"Who?"

"Mr. George Shiras."

"And anybody else?"

"I have known Mr. Shiras better than anyone else, but I have met various men from the Northern Peninsula."

"Do you remember anybody else that you know in Marquette County?"

"I know Mr. Hill."

"He is your attorney?"

"My attorney."

"You have known him since this case began?"

"No, I think I met him last year."

"Did you know him at the time?"

"There were several gentlemen got on board the train while I was passing through here."

"He may have been one of them?"

"I think he was one of them, sir, but I am not sure."

"That was your only acquaintance with Mr. Hill?"

"Yes."

"Mr. Shiras is really a resident of another state entirely and comes here sometimes in the summer, is that correct?"

"I don't know."

"But you don't understand that he makes his home here all the time, do you?"

"I couldn't answer that. I know that his grandfather and father were here, and that he has been here, but I don't know whether he lives here or not."

"Or whether he lives here in the wintertime?"

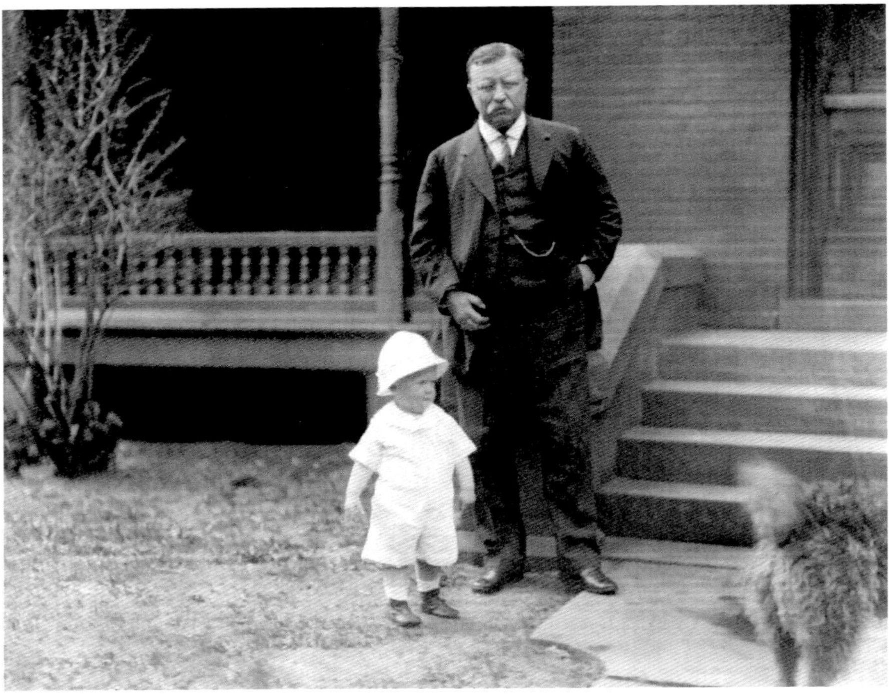

A rarely seen photo from George Shiras's personal scrapbook. Roosevelt is standing outside the Shiras home with the family dog and one of Shiras's children, a very candid personal moment between Roosevelt and the Shiras family. *Courtesy of the Marquette Regional History Center.*

"I know that he is off on hunting trips, natural history trips, a great deal, both summer and winter, and my understanding is that he is in Washington in the winter, but I may be mistaken."

"It was in Washington that you made his acquaintance?"

"I made his acquaintance I think when he was in Congress, but my chief friendship with him sprang from the fact that he is a student of natural history, and so am I."

"Mr. Shiras was in Congress from Pittsburgh?"

"In Congress from Pittsburgh."

"Do you know that that is his present home?"

"I don't know; I have been told so."

"During the time that you have related to the jury, your main occupation has been that connected in some way with politics, has it not?"

"At which time during the last fifteen years?"

"During the time you specifically related to the jury, beginning with your career in the legislature?"

"Beginning with my career in the legislature, I should say I was as much a writer and historian as I was a public servant, until I went to the Spanish War, but while I was governor and president, of course, those were my main occupation."

"Since that time, since your return from Africa, you have been engaged with the *Outlook*, that is a publication in New York City?"

"I have."

"And somewhat in politics also?"

"And in politics also, and might I amend my former answer. As I explained, I was for about thirteen years a great deal on my cattle ranch in the West."

"That was prior to the time I speak about?"

"Well, it was subsequent to my being in the legislature."

"It was?"

"It was. It began the last year I was in the legislature."

"Now, your acquaintances and vast number of associates while you were in Washington on business and in the city of Albany, were men also engaged in politics?"

"A good many of them were, and also practically every historian and naturalist to be found in the neighborhood of either Albany or Washington."

"You had among your acquaintances and associates a great many senators, did you not?"

"I knew almost every member of the Senate of the United States."

"You received them in the White House and were invited to their homes?"

"I received them in the White House. Only on rare occasions did they invite me to their homes."

"You knew also and associated with and received a great many congressmen. Did you not?"

"I received all the congressmen that came."

"And on your trips, you were received by them and entertained more or less?"

"Sometimes."

"On most of the trips?"

"Entertained by them, no."

A chuckle was heard from several attendees in the courtroom.

Andrews tried again: "You would be met by them?"

"I would be met by them on the trips while I was president; congressmen, senators, governors, private citizens, all met me."

"You were banqueted on those trips frequently?"

"It would depend upon what you mean by frequently."

"You went to various banquets?"

"Yes."

"On nearly all of those speaking trips that you speak of?"

"On almost all the speaking trips while I was president there would be public banquets."

"Now, since that time, you have been entertained in various homes of public men during the trips, such, for instance, as the home that you mentioned in Toledo?"

"Mr. Shepley, I don't think you would call him a public man."

"Well, at other places, you have been entertained by public men on these trips?"

"Only rarely; I was much more apt to be entertained by private citizens."

"Do I understand you to say that during all that time you never took anything in the way of intoxicants?"

"What do you say, sir?"

"If you took anything in the way of intoxicants, it was mainly a little brandy and milk?"

"No, you didn't understand me to say that. You understood me to say that about spirituous liquors."

"I will continue it to spirituous liquors. The only spirituous liquors you took in the main, that is in general, would be a little brandy mixed with milk?"

"A teaspoonful in a tumbler of milk, at night."

"And nothing else?"

"Nothing else."

"And that you never, in fifteen years, drank more than a dozen glasses of whiskey?"

"Oh, I didn't drink a dozen glasses of whiskey in fifteen years."

"Perhaps I have exaggerated it?"

"Yes."

"You never drank any whiskey, then?"

"No, I have said that in the fifteen years, outside of the teaspoonful of brandy that I have given you, and outside of the mint juleps of which I have spoken, it would be probably an overstatement to say that a dozen times I have drunk whiskey on the occasion of having a fever attack or some occasion like that, but I never drank a glass of whiskey on those occasions. The doctor would give me usually a little bit in a little tumbler with the lines graduated on it."

"A little graduated glass?"

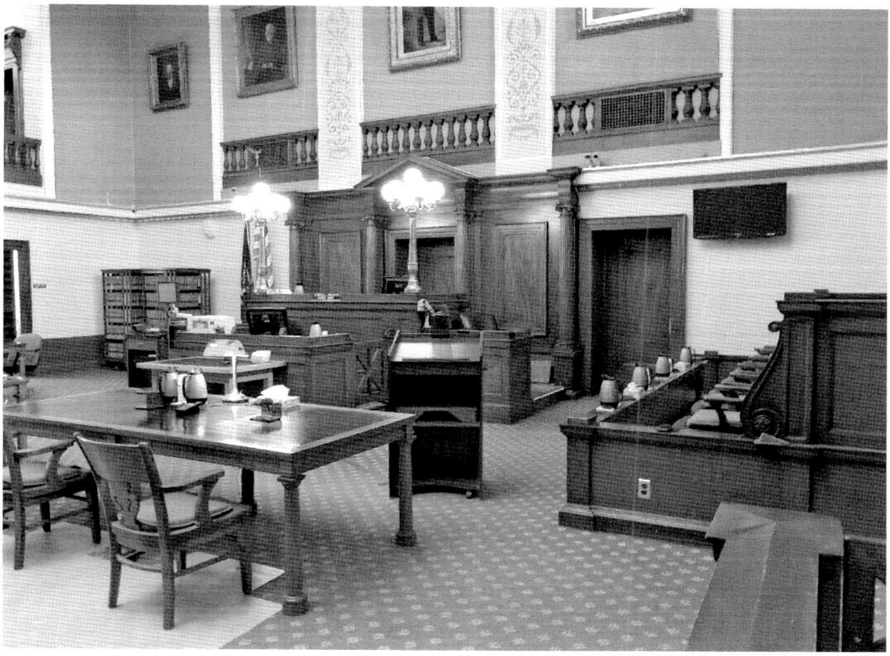

The interior of the courtroom remains the same as well. This shows the judge's bench, the jury box and the table where Newett and his lawyers sat. *Photo by Mikel B. Classen.*

"Yes."

"In that perhaps a teaspoonful or something like that?"

"A small amount, I couldn't tell you how much."

"And that you never took any whiskey or spirituous liquors except on the prescription of a doctor?"

"Excepting as I have described, I never did."

"And you never, as I understand you, drank a highball?"

"No."

"And you never have drunk more than five or six glasses of mint juleps in a year?"

"I have said I didn't average as many as five or six glasses in a year. I don't believe I drank as many as half a dozen glasses in a year, and you know those aren't tumblers, those are small glasses of mint julep."

"I don't know about the glasses you speak of. Do you understand that mint julep is made of whiskey?"

"So I understand, made either of whiskey or brandy."

"Now, your main drink, if you drink anything, is a little light wine?"

"The main spirituous, if I drink anything, is a little white wine."

"By light wine, what do you mean?"

"I mean so-called white wine, California white wine, or Sauterne, and sometimes at home a couple of wine glasses of Madeira instead of white wine."

"It would be either white wine or Madeira wine?"

"Yes."

"Any other kinds of wines?"

"I may have drunk a glass of sherry occasionally, but very occasional."

"Any others?"

"As I have described, at banquets or at the big dinners."

"I am asking about light wines."

"No."

"No red wine?"

"I don't drink red wine."

"You don't drink any red wine?"

"I wouldn't say I never drank any red wine. I may have been at some place where they passed around a glass of red wine and I may have touched it to my lips, but I don't like it. I don't drink it."

"You don't keep any wine of that kind in your house, do you?"

"Red wine?"

"Do you keep wine in your house?"

"Yes."

"Do you keep brandy?"

"Brandy and whiskey both."

"All right in your home?"

"I keep it at the house."

"Did you keep brandy and whiskey and wine and all these liquors in the White House?"

"I did; I inherited President McKinley's cellar too."

Andrews protested Roosevelt's answer: "I ask the court to strike out that which is not responsive to the question asked."

Judge Flannigan let him have the request: "You may strike out that part of the question."

It was Attorney Pound's turn to object: "Note an exception, please. I think he may state that," he continued. "I think it is competent for the witness to say that he continued the system that he found."

"If the witness answers that way, that he continued the system in the White House, it will be received," Judge Flannigan ruled.

"May I amend my answer?" requested Roosevelt.

"If you cannot answer directly, amend it any way you want to," advised Attorney Andrews.

"I can answer it directly, certainly."

Andrews was moving on: "I ask the question—"

Attorney Pound jumped in: "Wait a minute, let him answer it!"

Andrews continued, maintaining his demeanor: "I simply ask, did you keep liquors in the White House while you were president?"

"I continued the custom I found in the White House and kept liquors in the White House," Roosevelt finally got to reply.

"Did you keep two butlers to take care of liquor in the White House?"

"Not to take care of the liquor in the White House. They did that incidentally, as they had done it under the previous administration."

Again, Andrews protested Roosevelt's answer. "I ask that that be stricken out as not responsive to any question I asked him."

Flannigan, wanting to keep the proceedings moving, replied, "That answer may stand."

Andrews moved on. "Now, you frequently go to banquets?"

"Yes."

"You drink wine and champagne there?"

"I drink champagne or white wine. I prefer light wine, white wine."

"Or both?"

"Very rarely."

"You do sometimes?"

"I doubt if I do, but it would be possible on some occasions that I have drunk both."

"Do you drink brandy at the banquets?"

"No."

"Have you ever drunk brandy at a banquet?"

"I never have."

"Nor anywhere else?"

"Except as I have described."

"A teaspoonful in a glass of milk?"

"A teaspoonful in a glass of milk."

"Do you know Scotch whiskey?"

"I have seen it."

"Do you keep it in your home?"

"That I could not tell you. I know we did keep both Scotch whiskey and rye, because the guests at my house, when they come, are asked at dinner whether they wish whiskey, and I know they used to be asked whether they wished Scotch or rye. Whether we have it or not, I do not know."

"I addition to keeping it, you offer and serve liquors in your home to your guests, do you?"

"I do."

"Mr. Roosevelt, has that been your custom a good while?"

"I think it has been my custom ever since I have had a home, but that I am not sure of. I'm the first—"

"I think that answers my question," interrupted Andrews.

Judge Flannigan looked at Roosevelt and said, "You may proceed." The judge was giving Roosevelt every opportunity to tell his side.

Roosevelt continued with his answer: "It is possible it has only been my custom since I was governor or president, in as much as I never drink whiskey or brandy, I cannot be sure of it. I think that we had whiskey or brandy in the house before I was governor or president. I know we have had it since."

"Is that all the answer you care to give?" jibed Andrews.

"That is the answer."

"When did you first meet Mr. Pound?"

"The first time I remember meeting him was last fall."

"Do you know a Mr. Wallace of Detroit?"

"Yes."

"What is his business?"

"I don't know."

"Who had charge of the beginning of this suit, if you know?"

"That I cannot tell you."

"Don't you even know who began this lawsuit?"

"I was in Mercy Hospital, laid up with the shooting, when the suit was begun. This libel was brought to my attention after I had been shot."

"Do you know?"

"I am telling you. You asked me if I didn't even know that. I am telling you why I didn't know."

"Do you know who began the suit?"

"I do not."

"Do you know that Mr. Wallace had charge of it?"

"I knew that he knew of it."

"Whether he had charge of the employment of counsel?"

"That I cannot say."

"Had you met Mr. Pound up to that time?"

"I may have met him. I may have met him before that time on a trip in Michigan."

"Have you any memory of it?"

"I don't remember whether I even met him or not. I met hundreds of people then."

"Was Mr. Pound employed by you?"

"Is he employed by me now?"

"Was he originally employed by you?"

"You mean personally?"

"Yes, sir."

"In person, no. I was in Mercy Hospital on my back."

"I simply ask you whether you personally employed Mr. Pound. Can you answer that directly?"

"Of course it is possible that they asked me. It was by my direction that he was employed, but I didn't see him personally."

"Did Mr. Wallace have charge of the beginning of this suit?"

"Not that I know of."

"You don't know of any connection he had with it?"

"I know he was consulted about it."

"There were printed in the newspapers interviews in which Mr. Wallace stated that he had charge of the case and that the Progressive organization of this state had charge of it. Have you seen those interviews?"

"I have not."

"Have you borne the expense of the matter of taking testimony and all that kind of thing in this case yourself?"

"Yes."

"You have?"

"Yes."

"And it has not been done under the direction of the organization here in Michigan?"

"It has not been."

"Have you, in person, had charge of it?"

"Mr. Pound has had charge of it."

"Just Mr. Pound?"

"Yes."

"I think that is all, Mr. Roosevelt."

With that, Attorney Andrews ended his cross-examination. Attorney Andrews had succeeded in making Roosevelt seem confused about the facts, especially in regard to the origin of the lawsuit. Attorney Pound realized this and decided to do a redirect examination of Roosevelt. In later years, this confusion would lead to many errors on the origin of the libel suit. Since the libel suit was started on the day Roosevelt was shot, many would believe that Roosevelt started the lawsuit while recovering from the wound. Roosevelt's testimony reinforced that belief.

Attorney Pound began questioning Roosevelt after Andrews had finished. "As I understand it, you have no recollection of having seen Mr. Pound either in Milwaukee or Chicago prior to the beginning of this suit, which was sometime in October?"

"I have no recollection of it, Mr. Pound."

"As a matter of fact, you communicated, when you had sufficiently recovered, your desire directly to Mr. Pound by letter, did you not?"

"I did."

Andrews couldn't take it. "That is objected to as leading and incompetent. The letter is the best evidence."

Judge Flannigan reiterated the point. "The objection is that the letter would be the best evidence, Mr. Pound."

"That is right, and I do not propose to go into the contents of the letter. I am only wanting to know whether the witness assumes the responsibility of direction."

"I assume full responsibility of the direction, and no organization has any responsibility, excepting myself, or has paid, or will pay a dollar."

"Have you directed all the proceedings from that time up to the present, so far as the proceedings is concerned?"

"I have directed all proceedings. I have communicated with you and you have communicated with me direct. The only exception has been where I have communicated with Mr. Bowers and Mr. Van Benschoten as counsel in New York City."

"And the reason for that was, they were in New York, and you were not accessible to me?"

"The reason was, you were in Detroit. I wanted a lawyer that I could consult at once on any phase of the case that came up without your coming on from Detroit."

"And you were directed to get counsel close at hand?"

"I had to get counsel close at hand. I consulted you as to whom I should have."

With that, Roosevelt's testimony ended. T.R. had been on the stand for an hour and forty minutes. He stepped down from the witness stand and went back to the plaintiff's table. The courtroom was abuzz. The newsmen had been writing down everything they could. Roosevelt's testimony would be big news.

Next was the crux of Roosevelt and Attorney Pound's plan. There would be a parade of character witnesses that was intended to bury Newett and his claims stated in the editorial. Roosevelt wanted the rumors not simply stopped; he wanted them buried with a grave marker on them.

6

Roosevelt's Overwhelming Witnesses

When Roosevelt's testimony was over, Judge Flannigan declared a recess. Roosevelt went over to a table near where the news reporters were and sat on it, his legs dangling. The reporters wasted no time in pouncing on the opportunity. Roosevelt's little stunt with showing off his wound scar had gotten their attention.

A newsman jumped in with the first question: "Colonel, we never heard you were wounded."

"Well," began Roosevelt, "I wasn't officially wounded. I never went to an aid station. Somebody wrapped a bandage around my arm and I went on."

Roosevelt clearly was enjoying the attention. Dealing with the reporters was different than when he was campaigning. The newsmen weren't as adversarial. They were all looking for copy, and he was willing to oblige.

He continued, "You know, because of high blood pressure, I bleed more than most people. By the time the battle ended, I was drenched with blood." Roosevelt explained that the bleeding tendency was a part of his family, an inherited trait. He began another story. Once annoyed by its squeaking, he had climbed a sixty-foot windmill on his farm to oil it. At the onset, he had neglected to lock the wheel. The wind caught it, and a paddle sliced his scalp as he stood on the machinery platform.

"Big and heavy as I am," he explained, "it took me a long time to climb down. As I came to the back door, covered with blood, Mrs. Roosevelt met me. 'Theodore,' she said, 'I wish you would do your bleeding in the bathroom. You're ruining every rug in the house.'"

Roosevelt knew what he was doing. The press was going to write everything they could about the trial, and the stories would hit every paper across the country and internationally. He wasn't running for office now; it was his show.

The court went back in session with more witnesses testifying on behalf of Roosevelt. It was time for those who had traveled with him to speak on his behalf, and speak they did. For three days, they would speak.

After Roosevelt testified, it was Jacob A. Riis. He was a journalist, reformer and author of a book called *How the Other Half Lives*. He had become acquainted with Roosevelt during his time as a police reporter in New York City. The pair worked closely, trying to improve living conditions in the inner-city tenements.

Riis was asked by Andrews, "During the fifteen years you have known Mr. Roosevelt, did you ever see him under the influence?"

"Oh Lord, no," Riis responded. "The statement that he is a free user of liquor is a lie. I have seen the man under the greatest stress and never have I seen him resort to liquor."

"Is he a blasphemous man?" asked Andrews.

"No, he is a gentleman. About the worst I ever heard him say was 'By Godfrey!'" was Riis's answer.

Andrews protested, "I object. To testify that Roosevelt is a gentleman is incompetent."

Judge Flannigan ruled, "The law presumes the plaintiff is a gentleman."

DOCTORS ALEXANDER LAMBERT and Presley Rixey also testified. Rixey had tended Roosevelt during his time in the White House, and Lambert had been Roosevelt's longtime family physician and was an expert in the study of alcoholism. Lambert had also attended Roosevelt after the assassination attempt.

Roosevelt's attorney, Pound, made a point of bringing up the shooting: "When did your treatment begin, before or after he was removed to New York City?"

"I went out there and attended him in Chicago, the day after he was shot. Then I took him on to New York and treated him as long as it was necessary, and since then, he has been in good health, perfect health."

"At that time, the bullet wound was where?"

Andrews immediately objected, but Judge Flannigan allowed Pound to continue.

Dr. Lambert continued, "It struck him on the right side of the big chest muscle about an inch below the right nipple and went up and in four and a half inches, crossing over the sixth and fifth and breaking the fourth rib. It now lies one inch in and one-half inch down from the nipple."

"Do I understand you that he has made a complete recovery as far as possible under the circumstances?"

"Absolutely."

"As a professional man, with your acquaintance with Theodore Roosevelt, what do you attribute his recovery to?"

Again, Andrews objected, but Judge Flannigan commented, "It would depend upon the answer."

"I attribute it to his splendid unpoisoned physique!" Lambert answered.

"When he was president of the United States, did you have occasion to see him?"

"I frequently saw him, yes. I would spend days at a time and weeks at a time caring for his children when they were ill. I came and went from the White House very frequently."

"Have you eaten at his table?"

"Very often."

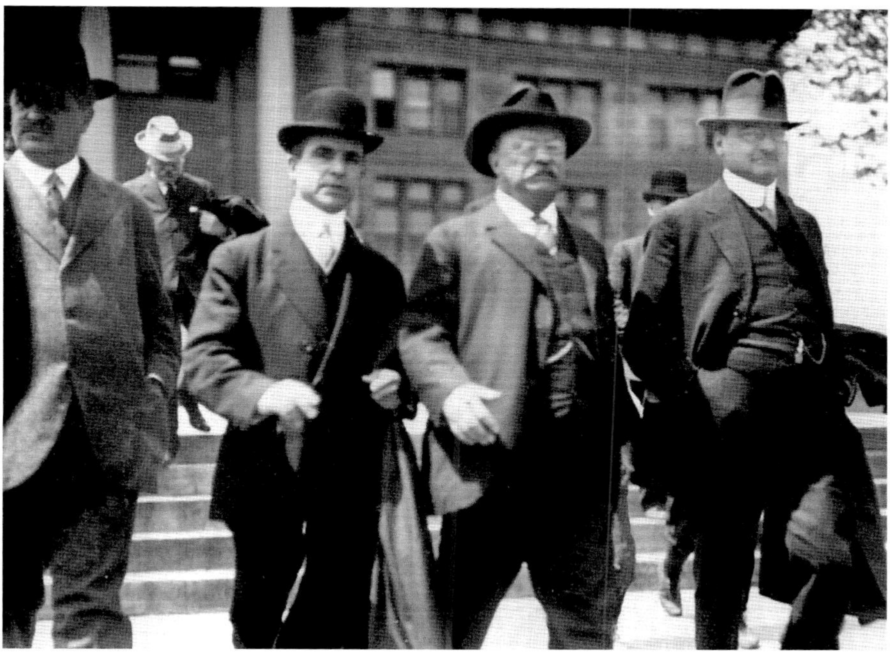

Roosevelt leaving the courthouse with some of the witnesses for the defense. *Left to right*: Frank Tyree, John O'Laughlin, Theodore Roosevelt and William Hart. This is another picture from the George Shiras scrapbook. *Courtesy of Marquette Regional History Center.*

"I ask you, from your twenty-two years experience, to describe to this jury from your observation of Mr. Roosevelt, what indulgence you have observed in alcoholic of malt liquors."

"He is an exceedingly temperate man, an unusually abstinent man."

"I will ask you, if he was a constant user to excess of alcoholic liquors when he was shot, would he have made the recovery he did?"

"No."

"Will you kindly tell this jury, in your twenty-two years' experience and your acquaintanceship with Theodore Roosevelt, as to whether you have observed any evidence of an alcoholic breath or anything of that kind upon the plaintiff in this case?"

"No, I never have, in all the twenty-two years I have been with him."

AT THE END OF THE first day, the news of Roosevelt's testimony went out. Headlines around the country proclaimed, "Roosevelt Tells Jury He Has Dislike for Whiskey and Brandy!" and "Roosevelt Reveals Secret Mint Patch at White House!"

The *St. Louis Dispatch* couldn't help but have a little fun with the trial. Since Roosevelt had testified that one of the only mint juleps he had away from the White House was at the St. Louis Country Club and that he had taken only one drink and passed it on was too incredulous for them, essentially saying he had perjured himself. According to the paper, the mint juleps made by the then-bartender, Tom Bullock, were just too good for anyone to taste and set aside: "To believe that a red-blooded man, and a true Colonel at that, ever stopped with just a part of one of those refreshments is to strain credulity too far."

An interview with former White House steward Henry Pinckney was printed. He worked at the White House from 1901 to 1909, and one of his duties was to be in charge of the juleps. According to him, when he sensed that Roosevelt had built up a thirst, he would go off to the mint patch, which was reported by the *Washington Post* to be on the White House grounds, "back of the executive offices, behind a lattice work house, where they hang the clothes to dry." Pinckney would start the julep by putting a few mint leaves in the bottom of a glass with a lump of sugar and a splash of water. "With a silver crusher I would bruise the leaves with the greatest of care." Once the sugar was dissolved into a nice minty syrup, he would add, not the bourbon found in a Kentucky-style julep, but a few drops of brandy and a "good slug" of rye whiskey of the finest kind procurable. Then Pinckney would fill the glass to its rim with cracked ice. But Pinckney had his own twist on the drink. He put the traditional sprig of mint so that the leaves dangled. Then, piled high on the ice, he arrayed "a bountiful supply of cherries, sliced pineapple, banana and orange." It was quite the fruity mix.

The juleps created such a sensation that the mint patch was added to the White House tour. In a wire service sidebar the next day appeared, "The White House mint patch, referred to by Roosevelt and now being used by President Wilson for garnishing spring lamb, promises to become as much an object of interest to capital visitors as the Washington monument. Already tourists are asking the White House police to point out the bed that gave the fragrant leaves for Roosevelt's occasional juleps."

The *London Times* published an article explaining to Brits American cocktails and how to make them.

As the day went on, more would testify on Roosevelt's behalf. The second day of trial presented a parade of witnesses testifying for T.R. First on hand was Washington correspondent John Callan O'Laughlin. O'Laughlin had been associated with Roosevelt for nearly twenty years. They met on the Nile in Egypt after the Africa expedition, and O'Laughlin traveled with Roosevelt to Paris as his personal secretary. O'Laughlin had also traveled with Roosevelt on some of his campaign trips.

Attorney Belden, Newett's head attorney, took over the cross-examinations. "During your twenty years' experience with Roosevelt, did you ever see him under the influence of liquor?"

O'Laughlin responded tersely, "It seems silly to me to have anyone suggest he was ever."

"At a banquet, did you ever see him drink liquor?"

"He sometimes took a glass of white wine. I never saw him drink more than one glass of champagne."

O'Laughlin went on to describe various dinners, official and otherwise, given in Washington, including dinners at the Gridiron Club. Asked if Roosevelt mixed his drinks at all or not, O'Laughlin answered, "If he took champagne, he never took anything else."

Attorney Belden asked if, in 1912 and for a time prior, there was not a general report among newspapermen that Roosevelt drank to excess.

O'Laughlin, shaking a finger at Belden, replied, "There is not a reputable correspondent in Washington but who thought the report silly."

The counsels on both sides were on their feet at this answer. The jury was ordered to exclude the answer by Judge Flannigan. Attorney Pound said the answer was "embarrassing," and Belden explained to the judge that he intended to show what other newspapermen had been writing on the subject of Roosevelt's alleged intoxication.

The argument began a lengthy discussion on the admissibility of "general" reports in the trial that Roosevelt used liquor to excess. Attorney Belden stated that it was the intention of the defense to question future witnesses using the line of defense, "that there was in existence a general report, amounting to a general reputation that Roosevelt does use liquors, and sometimes to excess; that reports published in newspapers of this charge never had been contradicted by him and that the defendant published a report of the former president's reputed excessive use of intoxicants, believing it to be true and not with any malice."

Belden held that such testimony was required to show good faith and absence of malice in the publication of the defendant's editorial. "It could not be in the presence of widespread comment by the public and important newspapers, that Roosevelt had been greatly damaged by a repetition of charges of intoxication in a remote weekly with a circulation of about three thousand copies."

Belden continued, "We propose to show such charges as made by the *Iron Ore* were by no means confined to this region. The plaintiff, during 1912, was the most talked of man in the United States, and we expect to introduce testimony to show adverse comment as to his habits were made in many parts of the country. This point bears strong in showing the absence of malice on part of the defendant. In the declaration to this the said plaintiff not only included the question of his use or non-use of liquors, but also that of his general reputation."

Judge Flannigan asked Belden whether the defense would fight the position of the plaintiff that malice in the publication of the article was the basis for damages.

Attorney Belden responded that proof of malice would undoubtedly increase the basis for damages, and the defense would argue against that assertion.

Attorney Pound broke in and made a point of clarification: "First I wanted to sue for fifty thousand, but Roosevelt insisted the amount should be moderate. He did not want to be vindictive and wanted merely damages for falsehood."

Belden continued his argument that precautionary damages in the case, under the law, might be any sum from six cents to $60,000, regardless of the amount asked in the lawsuit. Whether the alleged libel was true or not, the defendant believed the article to be true and did not publish it out of any malice.

Attorney Van Benschoten, one of Roosevelt's legal team, said that he wouldn't oppose the argument of the defense that the admission of evidence

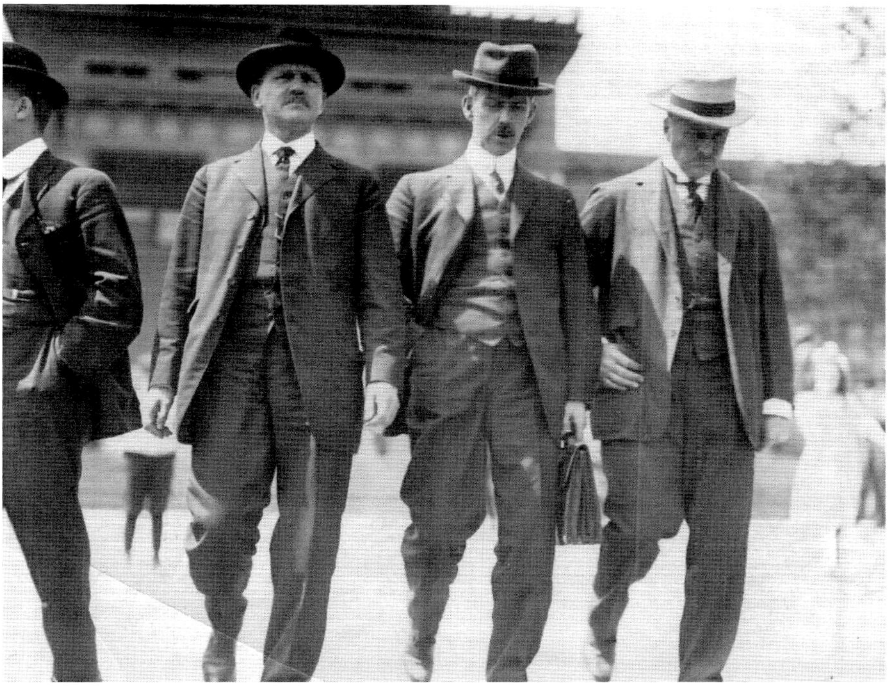

The Roosevelt trial lawyers exit the courthouse after a long day of testimony. In no particular order, they are James H. Pound, W.S. Hill and William H. Van Benschoten. *Courtesy of Marquette Regional History Center.*

bearing on Roosevelt's reputation as a liquor drinker was only to mitigate damages, not to prove intoxication.

Judge Flannigan called a recess. This called for some thoughtful decisions on his part. The court would reconvene at two o'clock.

When they did, the judge handed down two rulings. One would change everything for the defense.

First, the judge ruled that testimony by the defense touching on the general report said to have existed that Roosevelt used liquor to excess would be admitted. But the second ruling, that the defense would not be permitted to enter evidence that showed that other newspapers published similar charges relative to Roosevelt's use of liquor, was devastating to Newett's defense. They would not be able to use any other newspaper articles alleging the same charges as "The Roosevelt Way" had. They could talk about them, but they couldn't show them.

The court resumed its session, and the witnesses for Roosevelt began again. Robert Bacon, former assistant secretary of state; Truman H. Newberry, former secretary of the navy; Lucius F. Curtis, a newspaper reporter; and Gilson Gardner, another news reporter, all took the witness stand. Everyone was asked essentially the same questions: "Have you ever known Mr. Roosevelt to be under the influence of liquor?" "Have you ever smelled liquor on Mr. Roosevelt's breath." All said no. None had ever witnessed any time where the ex–Rough Rider had been drunk.

The day ended with more testimony, all supporting Roosevelt's nearly nonexistent drinking habits.

The third day of the trial saw members of Roosevelt's family on the stand. Both W. Emlen Roosevelt and Philip Roosevelt took the stand in support of their famous relative.

W. Emlen Roosevelt, a first cousin of Roosevelt, testified as to his sobriety. For a direct link to T.R., few were closer. His testimony was revealing: "The father of Theodore Roosevelt and my father were brothers associated in business together, and there was a great deal of intimacy between families. We lived within a few blocks of one another in the city of New York and were constantly associated."

Attorney Pound was doing the questioning: "Did that continue substantially up to the time of his arriving at the age of twenty-one?"

"It did. We were brought closer together every year. Before he was twenty-one, we had our country residence within a few hundred yards of one another at Oyster Bay."

"While he was an assemblyman, did you see him frequently?"

"I did. I was with him in the first political organization in Morton Hall, which he joined after leaving college."

"That is, you were both members?"

"We were both members."

"What is the fact, whenever the plaintiff was in New York, either visiting or on business, or at his home, since that time to the present, what has been your association with him?"

"They have been most intimate. When he was in the city of New York, from 1873 for a number of years, we lived in adjoining houses, and we were in and out of one another's houses daily. In the country, we had our homes within a few hundred yards, and we spent most of our days together. As we grew a little older, and he began his hunting trips, I joined him and was often with him both night and day for weeks at a time, sleeping in the same tent and doing the same work. When he became police commissioner, I saw a great deal of him. I went down to Montauk Point and was with him when he returned from the Spanish War. I was with him two or three times when he was governor of the state of New York at the executive mansion in Albany. I was with him when he was sworn in as vice president of the United States, and I was with him when he was sworn as president of the United States in the fall after the death of President McKinley. I was staying in the White House with him when he was sworn in as president after having been elected president of the United States. I stayed with him in Washington at the White House several days each winter. I was with him at a great many public functions, dinners and receptions. I was with him at the time that he had the peace meeting of the Japanese and Russians and made several trips with him of several days' duration on the *Mayflower*."

"The *Mayflower* was what kind of a boat?"

"The *Mayflower* was a government boat put at his service. I have been in a position to know every detail of his public and private life ever since he was a boy."

"I think impliedly you have covered it, but I want to ask you whether you have had meals at his house and he at your house, and if so, about how frequently?"

"It would be difficult to state how frequently, as it was our constant custom to be in one another's house at meals, as well as attending dinners by invitation."

"Was his house used as freely as your own?"

"It was. I would go in and out day or night."

"Tell the jury if there is any other man on earth that you have been as intimate with as with him?"

Andrews objected, and the judge upheld the objection. Pound had to rephrase his question. "I ask you this question, from your intimacy and your knowledge of the plaintiff, by association, by sight, by the exercise of every sense you have, of smell or other method, what can you say to this jury as to the habits of the plaintiff in relation to sobriety?"

"I will swear that never while I have been with Theodore Roosevelt was it possible for me to believe…"

"I object to that!" Andrews jumped up.

Emlen Roosevelt continued, "…it possible that he was in any way under the influence of wine or liquor of any character."

"Have you ever seen him under the influence of a narcotic or anything of that kind?"

"Never."

"What has been your business during your life?"

"I am a banker."

"Connected with what bank?"

"I am a banker of the firm of Roosevelt and Son, and Mr. Theodore Roosevelt's father and my father and my grandfather, my great-grandfather and my great-great-grandfather were members of that firm."

"I ask you, from your knowledge, what can you say as to the habits of the plaintiff as to profanity?

"I can say that he has one of the cleanest mouths—"

"Does he indulge in it or not?"

"He does not."

Emlen Roosevelt was allowed to step down. Another Roosevelt, Philip, had also come along to testify. Attorney Pound led the questions for him as well: "Now, taking your recollection of the plaintiff in this suit from your earliest age, your close association with him, what can you say to this jury as to whether at any time you have ever perceived the slightest indication of any influence of liquor of any kind upon the plaintiff?"

"Of course not."

"As a matter of fact, what do you say about his habits of speech, as to whether he is careless, or does he indulge in either profanity or obscenity in your presence, did you ever hear it, yes or no?"

"No."

"What liquor have you seen him consume?"

"I have seen him take a milk punch, as he calls it, and I have seen him take, on one occasion, two glasses of champagne, and I have seen him drink white wine, and I think that is all."

Besides the two Roosevelts, the day was again filled with witnesses testifying for T.R. The jury listened in rapt attention to the testimony. Roosevelt hadn't yet written his autobiography, and many of the details of the trial weren't known about Roosevelt. In the end, many of the details didn't appear in his autobiography at all. The details coming out about his personal and professional life were completely unknown at the time. The tales of hunting trips, exploring the regions of darkest Africa, exploits in the Spanish-American War and stories of the American presidency were as remote to the citizens of Marquette County as the Upper Peninsula was to the rest of the world at the time.

The Roosevelt witnesses even went so far as to include one of Roosevelt's manservants who had served him for ten years, an African American named James Amos, who said, "I never yet have served him with more than one full glass of champagne." He went on to say that "the colonel never drank at family meals, and when sharing white wine with guests, he would spritz his own glass with Apollinaris water."

James R. Garfield, former secretary of the interior and son of the former President Garfield, said that "even after the hardest ride, Theodore Roosevelt never drank anything but tea."

Lymon Abbott, editor of the *Outlook*, testified that Roosevelt's "ordinary beverage is milk."

A.Z. Blair, a former Rough Rider and Ohio judge, testified that during the campaign in Cuba, Roosevelt drank only coffee or water.

There was even a deposition by Admiral George Dewey read to the jury: "I have noticed that all people at dinners, whether they drink anything or not, are more or less excited. I have seen teetotalers very excited, and I have seen Mr. Roosevelt at dinners where he would be full of spirits, full of life and animation. All who knew him knew his peculiarities in that respect."

Roosevelt and his legal team had orchestrated an overwhelming case. A *Milwaukee Journal* dispatch from Marquette stated, "The Colonel is chief of staff, major general, captain, and orderly sergeant in marshaling his

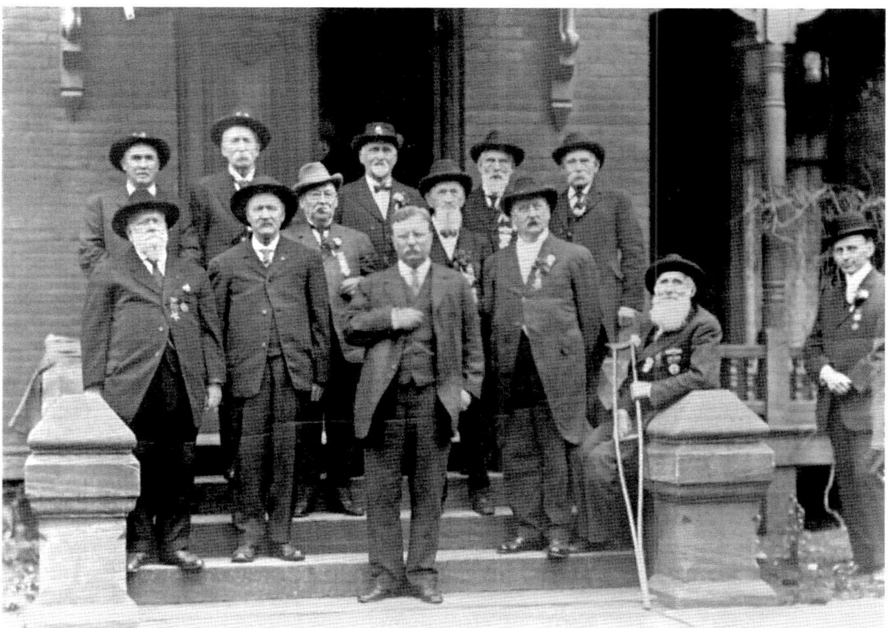

While in Marquette, Roosevelt paid his respect for the local war veterans with photo ops. This picture is with the Grand Army of the Republic (GAR) Civil War vets. *Courtesy of Jack Deo of Superior View Studio.*

evidence, leaning forward, his hands gripping the arms of his chair, his chin far out, he champs impatiently and fidgets."

At the end of the third day of testimony, the fourth day of the trial, it was Memorial Day. The trial was suspended for the holiday, and Roosevelt stayed at the Shiras home. He didn't participate in any of the local holiday activities. Reflecting on the day, he said sadly that it was the first Memorial Day he could remember in which he had not at least marched in a parade.

Roosevelt was asked to speak to a veterans' group, but he declined, not wanting his presence on the streets of Marquette to influence the trial in any way. Apparently some of the veterans weren't willing to take no for an answer. Veterans from the Civil War and the Spanish-American War made their way to the home of George Shiras, where Colonel Roosevelt was staying. They were welcomed, and Roosevelt took the opportunity to make an impromptu speech. Though there is no record of it, it was said to have been impassioned and inspired, a few reportedly moved to tears. Roosevelt also took some time to take photographs of himself and the veterans in front

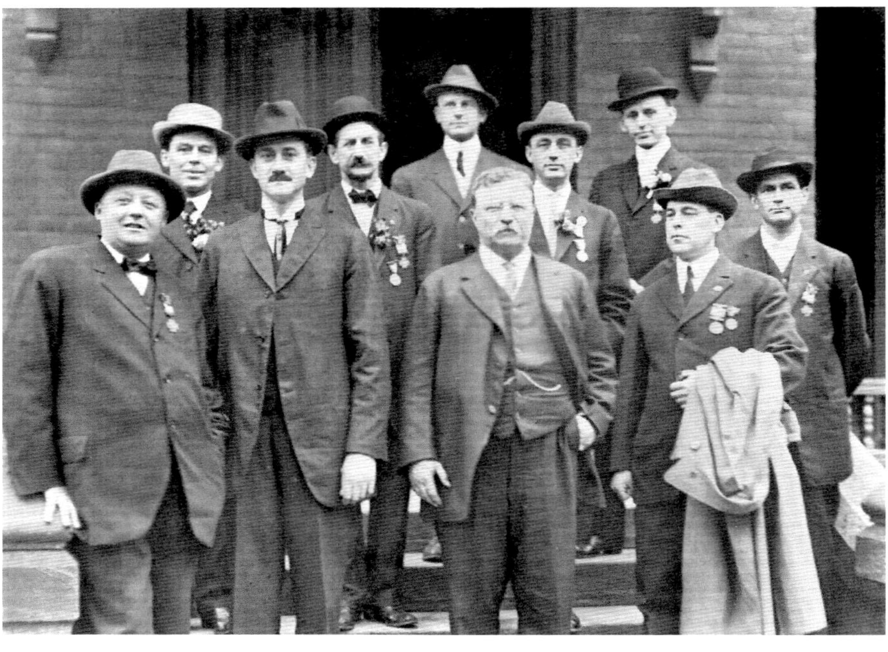

This picture has been often mislabeled. It has been identified as Roosevelt and his legal team and as the jury for the trial. All are incorrect. This is actually a picture of Roosevelt with Spanish-American War veterans of Marquette. Because of the mislabeling, their identities are unknown. *Courtesy of Jack Deo of Superior View Studio.*

of Shiras's house. One was taken with the Civil War vets and another with the Spanish-American War vets.

During the holiday, rumors were spreading around the city. It was speculated that the trial would last another week. The plaintiff's case was close to wrapping up, and the defense would begin. Many were saying that Newett had dozens of people locked away in area hotels waiting for their chance to testify against Roosevelt that he had been seen intoxicated on countless occasions. It was even rumored that a group of University of Michigan students were willing to swear that Roosevelt had been drunk in Ann Arbor during a talk he gave there in 1911. The next week of the case promised to be especially dramatic. The court was scheduled to reconvene the next day, Saturday. The trial was the topic of most conversations.

The *Grand Rapids Evening Press* got into the subject of Roosevelt's 1911 trip there by asking Senator William Alden Smith, who had been Roosevelt's host during his stay, about the drinking rumors. Smith said that Roosevelt had indeed been intemperate in Grand Rapids—while staying at the home of Mr. and Mrs. Edward Lowe, T.R. had used four lumps of sugar in his breakfast coffee.

Meanwhile, behind the scenes, a secretary for Roosevelt received a telegram asking for the arrest of James Miller, one of the witnesses Newett had interviewed and was to testify against Roosevelt, having claimed he'd witnessed Roosevelt drunk at Senator Joe Cannon's seventieth birthday party. The telegram stated that if he were to appear in Marquette, he was to be arrested. It further stated that Miller was wanted in New York on charges of grand larceny. It was signed J.P. Reynolds.

J.P. Reynolds, assistant district attorney, confirmed the report that he had asked for the arrest of James Martin Miller on the charges of grand larceny. Miller had written "checks full of bounce."

Supposedly, Miller had been friendly with Roosevelt at one time and had even written a book about him. He had apparently worked for Henry Ford on his newspaper, the *Dearborn Independent*. He was another person who had bad feelings about Roosevelt leaving the Republican Party. He had agreed to testify against Roosevelt.

Attorney Andrews, one of Newett's legal staff, made the announcement that Miller could not be brought to Marquette as a witness as "he is beyond Judge Flannigan's jurisdiction." The meaning of that quickly became evident.

The *New York Times* published a telegram from Miller hiding out in Winnipeg, Canada. Miller had skipped the country. It said that after he had made his deposition for Newett, "Roosevelt personally directed a campaign to destroy me."

It was a sign of things to come. When court would reconvene on Saturday morning, it would be a day of big surprises.

7

A Judicial Square Deal

Saturday morning brought excitement to those waiting inside and outside the Marquette County Courthouse. Roosevelt's lawyers were said to be wrapping up their part of the case that morning. It would be the defense's turn, and things were sure to get messy.

What they didn't know about was the quiet meeting that had gone on. James Garfield Jr. was approached by Attorney Andrews. He indicated that Newett might be willing to work a deal. Newett was realizing he was between a rock and a hard place. From all of the depositions that he had, no one was willing to take the witness stand. The one that had seemed serious, Miller, was now on the run from the law. Roosevelt's onslaught of highly credible witnesses had posed a juggernaut that had become insurmountable.

This turned into a serious offer from Newett. If Roosevelt would accept his deal, he would take the stand and admit the errors of his charges against Roosevelt if Roosevelt would agree to only nominal damages.

Roosevelt called a conference of friends and his legal staff to discuss the proposal. They advised him to take the offer. After all, he had accomplished what he set out to do. The testimony had overwhelmingly shown the truth about Roosevelt, and the trial had been covered internationally so the evidence of the trial had been published worldwide. No one would ever again be able to say Roosevelt had been a drinker.

They went to Judge Flannigan to get his opinion on the compromise. He was indignant. As far as he was concerned, Roosevelt had more than proven his side of the case and was entitled to all the damages he had sued for.

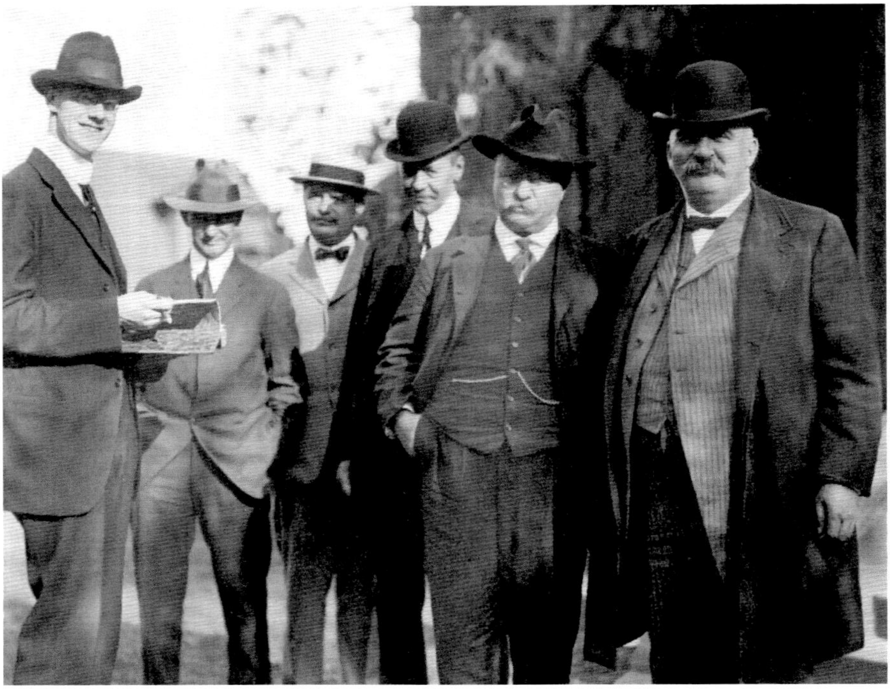

Roosevelt's legal team standing outside the Shiras/White house. Attorney James H. Pound stands to the right of Roosevelt. These lawyers do not look worried about the outcome of the case. This is another rare photo from the scrapbook of George Shiras 3d. *Courtesy of the Marquette Regional History Center.*

Flannigan felt Newett had been completely irresponsible with his editorial and, as a result, should have to pay for it.

When court was called to session on Saturday, the city was abuzz with anticipation. Everyone was certain the defense would begin before the day was out and that Newett's witnesses would begin breaking apart the testimony of Roosevelt and his string of witnesses and depositions. After all, everyone knew Roosevelt was a drunk, right?

The Roosevelt legal team wrapped up its defense of the former president. It entered several depositions, which were read into the record from the T.R. supporters who were unable to be there. Then the lawyers rested their case.

It was time for the defense. Newett would be the first to be called to the stand. He looked weak and sickly, his ailment appearing to get the better of him. His attorney, Belden, began the questioning.

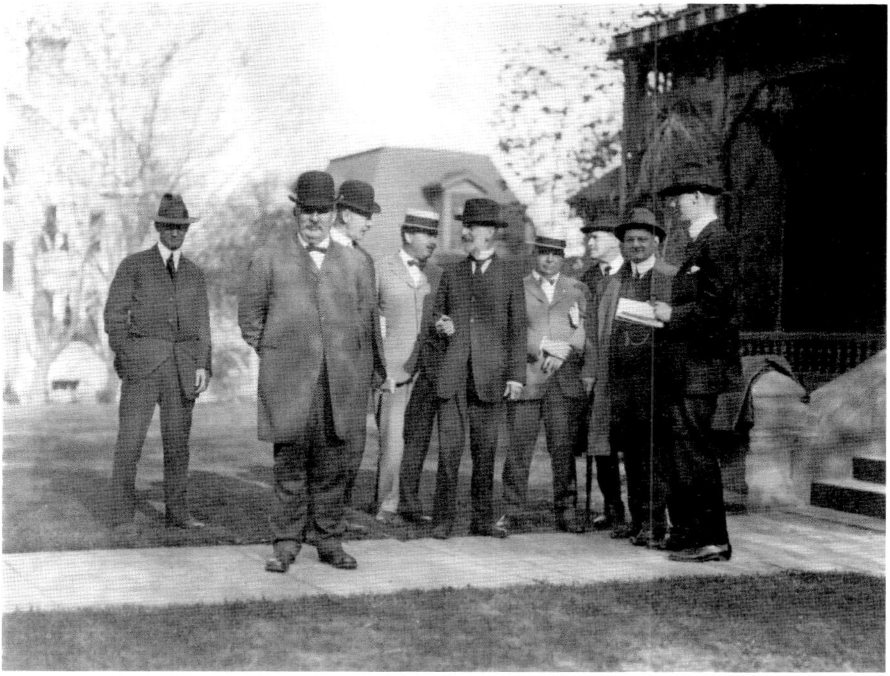

In another rare shot from the Shiras scrapbook, the Roosevelt entourage relaxes outside the Shiras house. Attorney James H. Pound is standing out front. The man in the center in the dark suit appears to be Judge Flannigan. *Courtesy of the Marquette Regional History Center.*

"When and where were you born?"

"On a farm near Janesville, Wisconsin, in 1856."

"How long did you live there?"

"Until I was sixteen years of age."

"How and in what manner did you gain your education?"

"In the district schools of that vicinity, attending in the winter and working on farms during the summer."

"When did you come to Marquette County to live?"

"In 1873."

"Have you lived here ever since?"

"I have lived here continuously ever since."

"When did you begin newspaper work, and on what paper?"

"In 1874 I began work on the first paper established in the city of Ishpeming, the *Iron Home*, and continued with the paper until 1879, when I engaged in business for myself on my own paper, the *Iron Ore*."

"Will you describe to the jury your weekly paper, the *Iron Ore*, what it covers?"

"It is a weekly paper, common to enterprising towns of this region, and specializes in mining news, news of local mines and also deals in news of mining camps throughout the country generally."

"In October 1912, what was its circulation?"

"From twenty-five hundred to three thousand, as near as I can recall."

"Over what territory was that principally distributed?"

"It circulated principally in Marquette County, probably four-fifths of the publication, and the balance largely in the Upper Peninsula."

"You may state whether or not you use wines or liquors yourself."

"I do not. I am a teetotaler and have very strong views on the subject; neither do I use tobacco in any form."

"You may state what offices you have held under the state administration."

"I was commissioner of mineral statistics from 1896 to 1900. I was appointed first by Governor Rich and my second term by Governor Pingree."

"State what offices, if any, you have held in the city of Ishpeming."

"I also held the position of postmaster from 1905 and held that for two years, at the end of which time I resigned. I received the appointment from the plaintiff in this case."

"Have you held any other offices?"

"I have held numerous offices in town, and now I am president of the Ishpeming Advancement Association."

Belden turned to the judge and requested, "May it please the court, the testimony in this case is of itself of such great importance that Mr. Newett has reduced his testimony to written form, and if counsel on the other side will permit me to give them a copy of this, I would like to have the witness go on and state in his own way the facts relating to this."

The judge ordered that it would be allowed. Newett took out his prepared testimony and placed it in front of him on the rail of the witness stand. He began:

> *Prior to 1912, I had been for many years a strong supporter of Theodore Roosevelt, recognizing him as a great Republican leader, and had frequently published editorials and other articles in the* Iron Ore *commending him and approving his policies, and I had therefore assisted in all his campaigns, not only by personal effort but also by financial contributions. Even in the primary campaign of 1912 I supported him as second choice for the Republican nomination. I mention these facts as indicating the impossibility of my harboring any feeling of personal malice against the plaintiff.*

Newett continued, to a hushed courtroom:

> *Sometime before 1912, I began to hear statements from various sources that Mr. Roosevelt was drinking wines and liquors to excess. During this period, I took a trip through the western country, passing through the states of Montana, Nevada and Arizona, as well as intervening cities. The statements as to Mr. Roosevelt's excessive drinking were repeatedly made in my hearing in various sections by many persons whom I believed to be reputable, but notwithstanding this I was loath to credit them. I felt that there was a possibility that these persons were mistaken and therefore thought the statements probably unwarranted. Late, during the winter of 1912, I took a trip to Florida, passing through Chicago and other cities en route, and spent several weeks at Belleair and other Florida points. During this trip, the same statements as to Mr. Roosevelt's habits were repeated in a very circumstantial way, although no one was able to say that he had actually seen Mr. Roosevelt drink to excess, or personally knew that he did so.*
>
> *During this period I also talked with various reputable people in Ishpeming, and elsewhere than the places before mentioned, who claimed to know the conditions as they existed in Washington and in whose statements I had the utmost confidence, both as to their reliability and their opportunity to know the facts. During the spring of 1912, newspapers came to me on our exchange list, various of which contained reference to Mr. Roosevelt's drinking habits. I saw and read these newspaper publications. From the statements so made in these papers and all the information which had come to me from the other sources mentioned, I felt that I could no longer doubt the truth of the statements which had been made, much as I regretted to believe them.*
>
> *When Mr. Roosevelt was nominated for the presidency on the National Progressive ticket, I opposed his election, having been a life-long Republican and believing that the success of that party would be for the best interests of the country.*
>
> *In October 1912, Mr. Roosevelt made a campaign trip through northern Michigan, and among other places, spoke at Marquette on the ninth. I was present on that occasion and heard him speak. In the course of his address he made what I considered a most unjust attack on our candidate for Congress, who was one of my life-long friends. This incident, together with the statements which had previously come to my knowledge, confirmed me in the opinion that Mr. Roosevelt ought not to be elected president, and I concluded that it was my duty in opposition to his candidacy to publish the*

statements which I then believed to be true. I therefore wrote and published the article which is here complained of. This publication was intended only as a blow to Mr. Roosevelt's candidacy for the position he then sought. In this publication, I acted in entire good faith, believing that the facts stated were true and believing that as a publisher I owed the duty to my readers to make that statement. After this article was published on the twelfth day of October 1912, neither the plaintiff nor any one in his behalf notified me that he claimed the charges so made were unfounded, nor did they request me to make any correction of the same. On the contrary, this suit was commenced October 25, 1912, and the service of the papers constituted the first intimation I received that the article was complained of.

After the commencement of this suit, there was nothing for me to do but to prepare to defend it, and I did so to the best of my ability. From that time, I proceeded to investigate the actual facts which could be shown by witnesses who would testify under oath. As I have said, up to the time of the publication, my information had been through persons who have claimed to have knowledge of the statements which had been made, and I went forward to verify those statements and determine the witnesses by whom they would be proven. Additional information came to me from various sections of the country, as this case had been given wide publicity. Both my attorneys and myself went forward with the investigation of all this with great thoroughness, and in numerous places in various parts of the country we found reputable witnesses who were willing to swear that from observation during certain addresses and public appearances of Mr. Roosevelt they believed he was intoxicated when they saw him. We have been unable, however, to locate or produce witnesses who will swear that they have actually seen Mr. Roosevelt drink to excess. Upon this phase of the case when the statements attributed to such persons have been sifted, it was found in each instance that the witness did not himself know that Mr. Roosevelt had drank to excess, or that if he had made such claim he was unwilling to testify. It is fair to the plaintiff to state that I have been unable to find in any section of the country any individual witness who is willing to state that he has personally seen Mr. Roosevelt drink to excess. I have taken the testimony in the form of depositions of more than forty reputable witnesses who have expressed the opinion that on those occasions as to which they testify he was intoxicated. I believe all these witnesses were honest in making their statements. I have relied upon those witnesses, but have recognized the lesser opportunities they have had to observe the plaintiff and his habits.

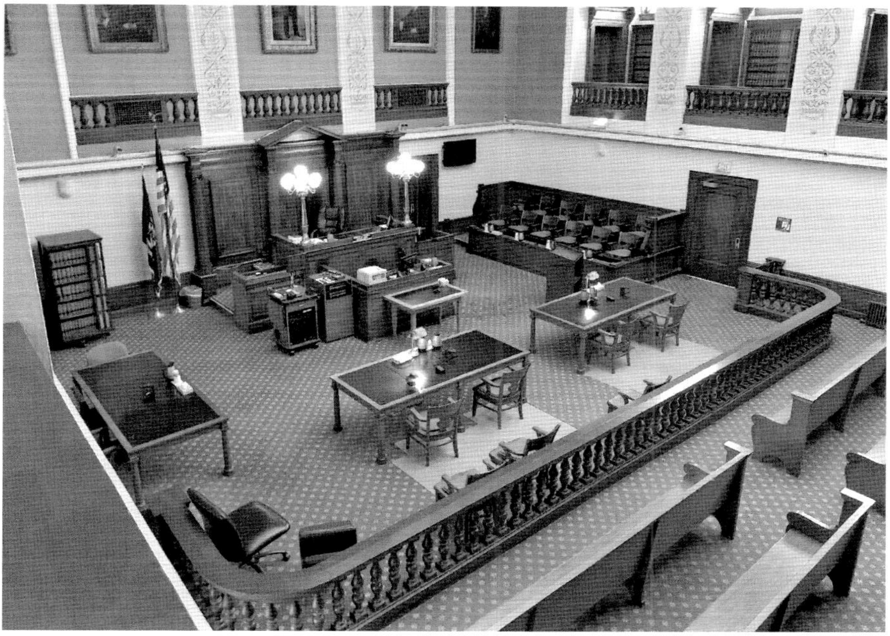

The view the society ladies of Marquette, Negaunee and Ishpeming had of the trial. The layout of the courtroom remains identical to that at the time of the Roosevelt trial. *Photo by Mikel B. Classen.*

The courtroom was shocked. Newett was giving up. There would be no defense. Roosevelt was smiling and fidgeting with his watch chain.

Newett continued reading his statement:

> *I have been profoundly impressed during the progress of this trial by the nature and extent of the evidence produced by the plaintiff to the extent that he did not in fact use liquor to excess on any occasion. I am unwilling to believe that these eminent men would purposely misstate the facts, or that under the circumstances related by them and their intimate acquaintance with the plaintiff for so many years they could be mistaken as to his habits. I have therefore been forced to believe that those who have given depositions or made the statement that in their opinion on the occasions to which they refer Mr. Roosevelt was intoxicated, had insufficient means and opportunity of correctly observing him and were mistaken.*
>
> *Up to the time of this trial I had believed that the statements in the article which I published were entirely warranted. But in the face of the unqualified testimony of so many distinguished men who have been in*

> *position for years to know the truth, I am forced to the conclusion that I was mistaken. I am unwilling longer to assert that Mr. Roosevelt actually and in fact drank to excess. As a publisher of a newspaper, I have never knowingly done injustice to any man, and neither I nor my attorneys are willing now to make or continue the assertion of an unjust charge against the plaintiff in this case. We have reached the conclusion that to continue expressly or implicitly to assert that Mr. Roosevelt drank to excess or actually became intoxicated as set forth in the article would do him an injustice. Since in publishing the article I acted honestly and in good faith, I propose at this time and throughout the remainder of this case to occupy a like position. My position throughout the introduction of my defense is and will be that in the publication I acted in good faith and without malice.*

Newett ended his statement and sat waiting for a response. The courtroom was stunned. Suddenly the trial was over—there would be no anticipated stories of Roosevelt's drunkenness or the wild misbehavior that the rumors had told.

Roosevelt's attorney Van Benschoten stood and spoke: "If the court please, the plaintiff has asked me to ask the court if he may have the privilege of making a statement."

Judge Flannigan agreed.

Theodore Roosevelt stood and spoke with a grin: "Your Honor, in view of the statement of the defendant, I ask the court to instruct the jury that I desire only nominal charges. I did not go into this suit for money. I did not go into it for any vindictive purpose. I went into it, and as the court has said, I made my reputation an issue, because I wished, once and for all during my lifetime, thoroughly and comprehensively to deal with these slanders, so that never again will it be possible for any man, in good faith, to repeat them. I have achieved my purpose, and I am content." With that Roosevelt sat down, his statement ended.

Attorney Belden then asked for a recess, which the judge granted. The attorneys and Judge Flannigan met during the recess. Judge Flannigan, who firmly believed Roosevelt had been wronged by Newett's editorial, again stated he believed Roosevelt deserved the full amount that had been asked for, $10,000. Roosevelt still wasn't interested in the money; he had only wanted the cessation of the continued assassination of his character. He had put the detractors and gossipmongers on notice. Roosevelt's legal "Big Stick" had struck a resounding blow.

As court resumed, Attorney Belden officially ended the defense's case: "In view of the statement made by the plaintiff in this case, we have determined that further testimony will be unnecessary, and we therefore rest our case."

Attorney Pound responded, "May it please Your Honor, not only upon the statement of the defendant himself, and the concession of the plaintiff, as to what he claims, I ask Your Honor to direct a verdict in this case in favor of the plaintiff, as a matter of law."

Judge Flannigan turned to the jury and addressed them:

> *Gentlemen of the Jury: the position which the plaintiff has now taken enables the court to dispose of this case finally at this time. This action, which is for libel, followed upon the publication by the defendant of the statement that the plaintiff was frequently drunk. Libel is a malicious publication expressed in printing or writing, or by signs and pictures tending to blacken the memory of one dead, or the reputation of one who is alive and expose him to public hatred, contempt and ridicule.*
>
> *In all actions for defamation, malice is an essential element. But in such actions, the word "malice" is understood as having two significations. One, its ordinary meaning of ill will against a person, and the other its legal signification, which is a wrongful act done intentionally without just cause or excuse. These distinctions have been denominated as malice in fact and malice in law. Malice in fact, or actual malice implies a desire to injure. Malice in law is not necessarily inconsistent with an honest purpose, but if false and defamatory statements are made concerning another without insufficient cause or excuse, they are legally malicious, and whenever the natural tendency of the publication is to impair the plaintiff's reputation and the publication is not privileged, malice is implied from the defamatory nature of the publication and its falsity.*
>
> *The plaintiff was formerly president of the United States. At the time of the publication, he was candidate for the office of president of the United States, and the defendant, by his plea, claimed that the plaintiff being such a candidate and he, the defendant, being the publisher of a newspaper, he was privileged to comment upon the official acts, character and conduct of the plaintiff; that the publication was made in good faith and with an honest purpose to enlighten the public upon the character and fitness of the plaintiff for the position he sought, and that therefore the publication was privileged.*
>
> *The publisher of a newspaper may freely discuss the fitness of a person for public office; he may lawfully communicate to the public any fact within his knowledge respecting the official acts, character or conduct of the candidate and freely express his opinions on and his inferences from*

such facts, so long as he states as facts only the truth, and as opinions and inferences therefrom only honest belief. To that extent, the publisher of a newspaper is within his privilege, and any candidate for public office injured in his reputation or feeling by reason of the publication of real facts, and of the publisher's opinions thereon, and inferences therefrom, honestly and in good faith entertained, has no redress.

But Michigan has not yet granted to anyone, whether newspaperman or not, the privilege of writing or speaking of a candidate for an elective office words which are both defamatory and untrue. Whatever the law may be in other jurisdictions, with us no one is privileged to defame a person because he happens at the time to be a candidate for public office. When a newspaper attacks the character of a candidate for public office, the publisher must, at his peril, be careful to ascertain the truth of the facts charged, as in any other case of libel. From what has been said, it follows that the defendant's liability in this action depends solely upon whether the charge is true or false.

Unwilling to rest on the presumption of law that the charge was false and called upon the defendant to prove its truth or answer in damages, as was the plaintiff's right, the plaintiff proceeded before this court to show that the charge was false in fact. And by his own, and the testimony of a long list of witnesses of high character and unimpeachable credibility, he has satisfied the defendant, the jury also, I fully believe; certainly he has convinced the court, not only that he never was drunk, but that he is now, and always has been an exceedingly temperate and abstentious man. After hearing the testimony of the plaintiff and his witnesses, the defendant, freely, in the presence of the court, conceded the mistake of fact in his publication. In this, he has taken a just position.

The court is satisfied from the testimony of the defendant that, in publishing the statement, he was not actuated by actual ill will towards the plaintiff and that he acted in good faith. The enjoyment of a private reputation unassailed is as much a constitutional right as the possession of life, liberty or property, and whenever one's reputation is unlawfully assailed, he is entitled to recover all the damages which he has actually sustained. The law presumes that damages must actually, proximately and necessarily result from the publication of defamatory matter whenever the immediate and actual tendency of the defamatory words is to impair the plaintiff's reputation. Such was the immediate and natural tendency of the publication of the words we are considering, conceding that the publication was made in good faith and in the absence of actual malice.

Therefore, accepting as true all which the defendant now claims, yet the injury to the reputation and feelings of the plaintiff which naturally, proximately and necessarily followed upon the false publication would warrant a verdict in the plaintiff's favor in a substantial amount, and would sustain a verdict in any sum up to the amount claimed in the plaintiff's declaration, which is $10,000.

But as the court is advised by the plaintiff, the object of the plaintiff in bringing and prosecuting this action being the vindication of his good name and reputation, and not the recovery of a money judgment, and he having in open court freely waived his right to the assessment of his actual damages, it only remains for the court to direct a verdict in his favor for nominal damages, which, under the law of Michigan, is the sum of six cents.

You are therefore, gentlemen, directed to render a verdict in favor of the plaintiff for that amount.

As Judge Flannigan finished his directions to the jury, Attorney Pound rose and made a request: "So that it may appear on the record, I formally ask Your Honor to enter an order for the suppression of the depositions not used."

This move by Attorney Pound would forever seal the depositions that Newett had collected, and no one would ever know who made them or see their content. The accusations were buried forever. Judge Flannigan didn't hesitate in ordering it.

An agreement by both parties tells of the fate of the depositions: "It is hereby stipulated between the above named parties by their respective attorneys, that all depositions except those of the plaintiff which were read into evidence in the trial of said cause, shall be withdrawn from the files of said cause and destroyed."

The jury retired.

8

The Price of a Good Paper

When the members of the jury came back, they rendered the decision that they'd been ordered to. The foreman, William Matthews, stood and read the verdict: "The said defendant is guilty in manner and form as the said plaintiff hath, in his declaration in this cause complained against him, and that they assess damages of the plaintiff on occasion of the premises at the sum of six cents."

Judge Flannigan then adjourned the court. Roosevelt immediately rose and went over to the jurors. He shook the hand of each one, taking a moment for each and saying as he went, "I thank you, gentlemen, each one of you. It was splendid, just perfectly splendid." The jury could now, finally, go home to their families. The bivouac at the Marquette County Courthouse was over.

Matthews said the jurors had been impressed with the testimony of each witness. "After Mr. Roosevelt's testimony, I think we were impressed chiefly with that of Mr. Garfield and Mr. Riis." He continued, "The statement of Mr. Newett himself that he could not doubt that Mr. Roosevelt was a temperate man after considering the character of the witnesses was just how we felt."

Roosevelt commented to a reporter about Judge Flannigan, "Judge Flannigan has given us a model of the conduct which should distinguish the conduct of the highest court in the land."

T.R. milled around the room talking to friends and supporters, shaking hands and continuing to compliment his attorneys and Judge Flannigan.

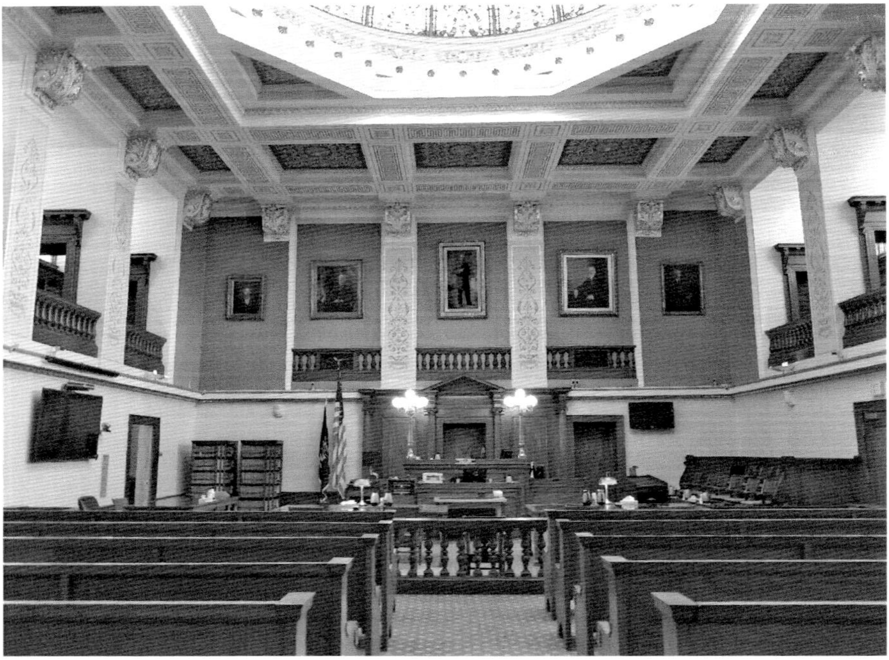

The Marquette County Courtroom, the view from the back of the gallery. The great domed stained-glass skylight can be seen in the top of the picture. *Photo by Mikel B. Classen.*

A reporter from the *New York Times* asked Roosevelt, "Are you and Newett going to meet?"

He answered, "Not if the advances are to come from me."

Newett, as well, became surrounded by friends and supporters. In his public admission of error, he had only strengthened his stature with community and friends. Continuing to appear ill, he was the first to leave the courtroom without meeting or speaking to his famous opponent.

As Roosevelt was leaving the courthouse—he wanted to get home as quickly as he could to possibly salvage a small piece of the holiday weekend with his family—a reporter asked him one last question: "What do you intend to do with your nickel and a penny?"

Roosevelt laughed, "Six cents is the price of a *good* newspaper." The *Iron Ore* sold for three cents.

The *Houghton Mining Gazette* made these observations upon the sudden ending of the trial:

> *There is not the slightest doubt that Colonel Roosevelt is a mightier and more potent political influence in this country as a result of this trial than ever he has been before. There is no doubt at all that a great many people honestly were of the opinion that they had seen the Colonel intoxicated upon numerous occasions. At the same time, in view of the testimony and in view of the verdict, that intoxication was the exuberance of high nervous tension of a naturally active man. To many who had never seen Mr. Roosevelt in action, his enthusiasm at the conclusion of the trial might easily be mistaken for the intoxication that comes from too liberal consumption of alcoholic stimulants. During the trial he exhibited nervousness all the time and his facial contortions were those of a highly strung individual to whom some kind of action is always necessary to give vent to unusual mental strain. The verdict is proof conclusive that the numerous tales which have been so generally circulated were based entirely upon erroneous premises, in many cases honestly expressed, but untrue, nevertheless.*

In an editorial dated June 4, 1913, the *Marquette Mining Journal* commented on the aftermath of the trial: "It has in many respects been unpleasant, but it has been salutary with Col. Roosevelt as its principal beneficiary. It closed in a fortunate manner, and with much less of ill feeling than it might have been expected to leave behind it. In Marquette County, as far as the *Mining Journal* can learn, there is practically none."

For those who witnessed the trial and were there trying to get a glimpse of the former president, Roosevelt had given the city of Marquette a great gift, a moment in history, one that wasn't lost on anyone. The city of Marquette would be forever linked to the Roosevelt legacy. The affection for Roosevelt lives on to this day.

W. Emlen Roosevelt published a transcript of the trial so that history would know the truth. In the foreword he writes:

> *To My Fellow Witnesses:*
> *We have all been greatly incensed from time to time in recent years, by reports that Theodore Roosevelt was a drinking man. We, who knew the unusual purity and wholesomeness of his life, felt especial anger at this rumor, both because of its persistence and of its cruel falsity.*
>
> *When Mr. Newett published the libel in the* Ishpeming Iron Ore *we had our opportunity to contradict it once and for all. We were glad to appear under oath and testify to Theodore Roosevelt's clean habit of life. We were glad to contradict this story completely and authoritatively. I have had this*

book printed because I knew that we all wanted to have a complete copy of the official record which contradicts the libel.

In a strange twist of fate, Roosevelt himself would be sued for libel just two years later. In Onondaga County, New York, in late May 1915, Roosevelt

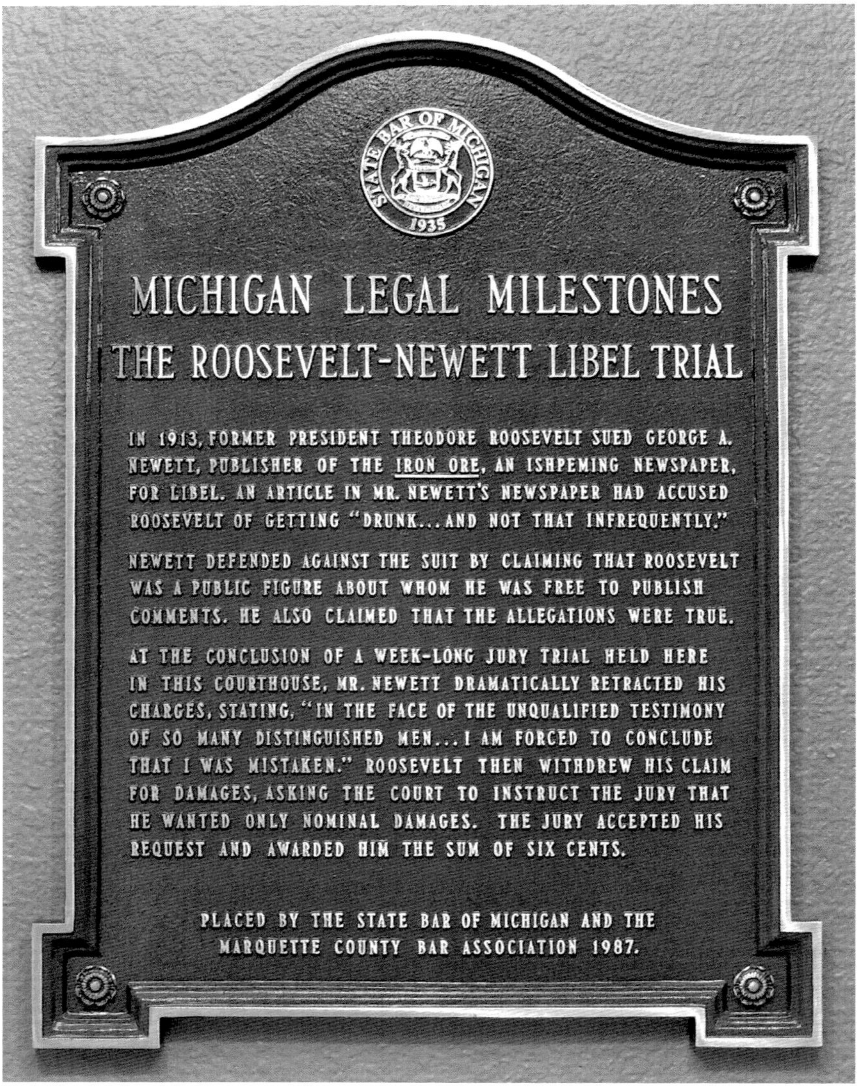

A plaque hangs outside the courtroom as a marker for the Roosevelt libel trial. Marquette County is proud of its history, and the courthouse is no different. *Photo by Mikel B. Classen.*

would accuse a William Barnes, a leader in the New York Republican Party, of being a party boss.

Roosevelt's comments, "The State Government [of New York] is rotten throughout in almost all its departments, and this is directly due to the dominance in politics of Mr. Murphy and his sub-bosses…aided and abetted where necessary by Mr. Barnes and his sub-bosses," would send him back to court, this time defending his own words.

Roosevelt would win this case too, assisted by the testimony of his nephew and cousin, assistant secretary of the navy, Franklin Delano Roosevelt.

Governor Fred Green appointed Judge Flannigan to the Michigan Supreme Court fourteen years later. His reputation as hardworking and straightforward but compassionate would stay with him throughout his life.

In 1987, a historical marker was placed in front of the Marquette County Courthouse making it an official historic landmark. The State Bar of Michigan donated the commemorative plaque in conjunction with Michigan's sesquicentennial. Along with this, a bronze plaque was placed on the wall outside the courtroom, further commemorating the trial. The courthouse would again come to national attention when it became the primary location for the film *Anatomy of a Murder*.

Epilogue

Teddy Has the Last Word

A few years after the libel trial, Roosevelt remarked on the incident. He was in his trophy room at Oyster Bay when one of his guests asked him, "Did you ever smoke?"

Roosevelt replied:

> *There is where that story of my drinking started. You see, when I would decline a cigar, saying I did not smoke, folks would often ask, in a joking way, "What are your bad habits?" In the same spirit, I would reply, "Prize fighting and strong drink."*
>
> *Now it so happens that the Lord in his infinite wisdom elected to create some persons with whom it is never safe to joke—solemn asses who lack a sense of humor. I am very fond of that story of Sidney Smith's, who, playing with his children, stopped suddenly, saying, "Children, we must now be serious—here comes a fool." You know the kind he meant—those poor unfortunates who must take everything said to them literally.*
>
> *One of these to whom I made that remark said, "Roosevelt, I hear, drinks hard." The other fool replied, "Yes, that's true. He told me so himself."*
>
> *And so the story went on its travels.*
>
> *That is all there ever was to the talk of my drinking. From that start, it spread until, in self-defense, I was compelled to take action to stop it. Some folks have said I went out of my way to find a little editor of small means and few resources of evidence who could not well defend himself. The fact is, he was the one editor I could hold to account. There were and are editors*

nearer New York I gladly would have sued under like circumstances, but they knew better than to print what they knew was untrue. Had any of them done so, I would have hauled them up short, and with much more glee than I did the Michigan man, for the men I have in mind have real malice toward me, and he, I am satisfied, had none.

We parted good friends. I certainly had nothing against him. In his zeal to do things, he put in print what shrewder and really malicious men who would harm me if they could, dare not print. I believe he was honestly sorry when he found his error.

However, the thing had its value. We're never too old to learn, and I learned to be careful with whom I cracked the simplest joke. Thank God, there are many you can joke with in safety. If we couldn't laugh once in a while, what a world this would be! It wouldn't be a world—it would be a madhouse.

Addendum

Shiras's Commentary on Roosevelt and Mint Julep Recipe

Sometimes there are things that are found in the research that don't specifically apply to the subject at hand but are too fascinating to ignore. One such thing concerning Roosevelt is a commentary by George Shiras 3^d^, Roosevelt's friend and associate in the subject of natural history. The pair became quite close over the years and often shared time and conversation in each other's company. They frequently exchanged letters in which Roosevelt was endlessly urging Shiras to write a book about his studies of the natural world.

Shiras considered Roosevelt a genius, but Roosevelt felt that the concept simply wasn't true. Roosevelt felt that the term simply didn't apply to him. The pair had a conversation about that very thing at Roosevelt's home in Oyster Bay.

Shiras himself tells the story:

> *On one occasion, I spent a night at Oyster Bay with Colonel Roosevelt, and in the evening, we sat on the side porch facing the sound. In the course of the talk, my host spoke disapprovingly of those who referred to him as a genius, a term he declared was most inapplicable and objectionable in his own case.*
>
> *"If there is one thing," he said, "that may have exemplified my career, it is what can be accomplished by most persons through hard work and making the best of every opportunity.*
>
> *"My rule," he remarked, "is a simple one. Do the best you can with what you have and do it now."*

An epigrammatical appeal for conscious work and a terse warning against the besetting sin of procrastination, I replied.

"Why," exclaimed the colonel, "I have been a second or third rater in everything I have tried to do, and any accomplishment must be credited to industry and not to any providential endowment arbitrarily separating me from the rest of the community."

Nevertheless, I remarked, you are a genius, and a conspicuous one at that. And thereupon, I endeavored to summarize his remarkable and varied achievements as best I could:

Author, editor, historian, lecturer, legislator, ranchman, sportsman, naturalist and explorer, athlete, police commissioner, and coworker with

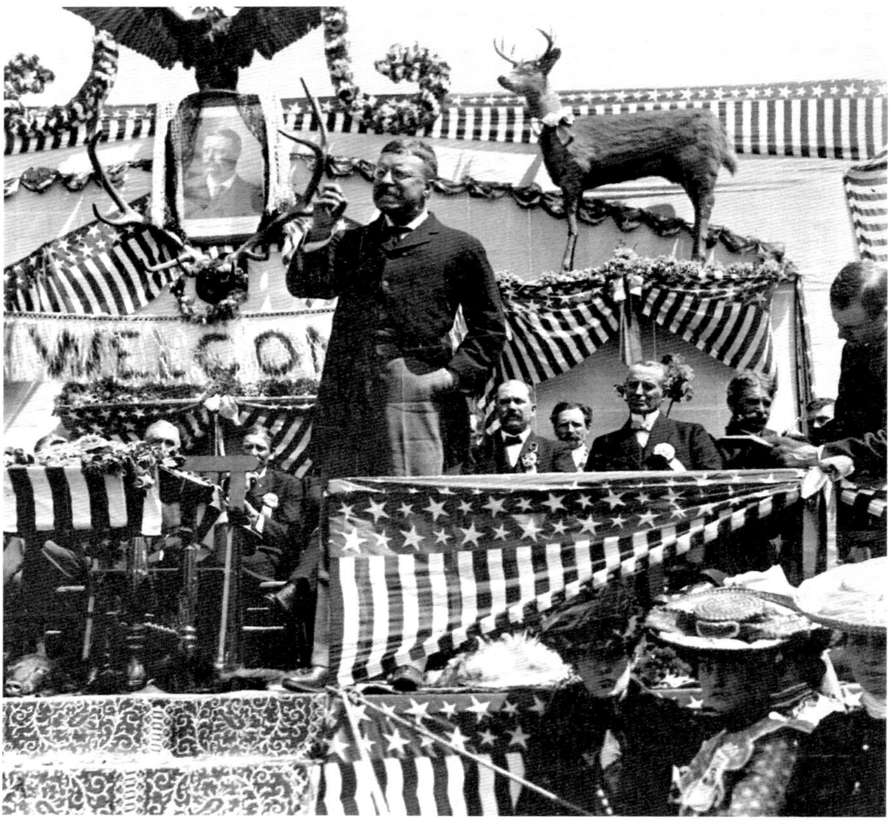

This picture is from an old stereoview of Roosevelt campaigning. It shows the intensity of Roosevelt while speech giving. Surrounded by a hunting motif, it is easy to see how he appealed to Upper Peninsula voters. The stuffed vulture perched atop Roosevelt's portrait presents a strange image. *Courtesy of Jack Deo of Superior View Studio.*

Jacob Riis in social betterment, rehabilitator of the navy, a successful military leader and a foremost promoter of international peace, the chief executive of a great state and of the nation, a conservator of natural resources, a statesman and active political leader, the foe alike of capitalistic tyranny or industrial demagoguery, a human link between all grades of mankind from kings to pugilists and, finally, one who, by his exemplary private life and moral teachings, made an impress on the country and the world in general quite as valuable as his successful discharge of many public duties.

In conclusion, I pointed out that while he may not have taken a first prize, allotted to only the specialist or one particularly endowed, he had taken so many second and third prizes that he far outpointed any competitor and therefore was, in fact, a genius, a genius in versatility, the greatest of all, from the standpoint of mentality or of public service.

"Well," declared the colonel, "I'm glad to see you admit that I never stood at the top of any one of the many things I endeavored to do, and I therefore accept your apology."

And then he chuckled in his inimitable way.

TEDDY ROOSEVELT'S MINT JULEP RECIPE

fresh mint
1 sugar cube
¼ oz brandy
4 oz rye whiskey
sliced pineapple, sliced banana, orange and cherries

Gently muddle a few leaves of mint with the sugar and a good splash of water in the bottom of a glass (or silver julep cup if you have one). Add brandy and whiskey, and then fill the cup to the rim with pulverized ice. Stir until the outside of the glass is thick with frost. Pile the top of the drink high with mint and fruit. Drink to one's heart's content.

References

Calumet (MI) News. "Both Roosevelt and Newett Win Important Points." May 28, 1913.

———. "Famous Libel Suit Started at Marquette." May 26, 1913.

———. "Newett Gives Up Fight and Col. Roosevelt Asks Only Nominal Charges." May 31, 1913.

———. "New Witnesses Testify as to T.R.'s Sobriety." May 29, 1913.

———. "Roosevelt Tells Jury He Has Dislike for Whiskey and Brandy." May 27, 1913.

Collier, Peter. *The Roosevelts: An American Saga*. New York: Simon & Schuster, 1995.

Detroit News. "Jay Hayden Recalls: Teddy Roosevelt's Day in a Michigan Court." March 28, 1965.

Grand Rapids (MI) Press. "When Teddy Took Marquette." January 26, 1969.

Houghton (MI) Mining Gazette. "Editor Newett's Statement Acknowledging He Erred in His Charge of Drunkenness." June 1, 1913.

———. "Newett Retracts His Charge of Drunkenness." June 1, 1913.

Leary, John J., Jr. *Talks with T.R.* New York: Houghton Mifflin Company, 1920.

Marquette (MI) Mining Journal. "Ceremony to Mark Roosevelt's Libel Suit." May 9, 1987.

———. "Half Century Has Gone by Since Roosevelt's Marquette Libel Suit." June 14, 1963.

———. "Iron Ore One of Earliest Weeklies in U.P., Goes to Press for Last Time." March 28, 1951.

———. "Jury in Case of Roosevelt vs Newett Filled Out Last Night." May 27, 1913.

———. "Roosevelt's Life Topic of Address." October 26, 1926.

———. "T.R." April 4, 1958.

Milwaukee (WI) Journal. "Teddy Roosevelt Had His Day in Court." November 4, 1968.

———. "When T.R. Sued to Prove Sobriety." June 13, 1953.

Mitchell, James. "Bull Moose Tracks in Marquette, Michigan." *Stereoscope: The Journal of the Historical Society of the United States District Court for the Western District of Michigan* (July 2012).

O'Toole, Patricia. *When Trumpets Call*. New York: Simon & Schuster, 2004.

Roberts, Dave. "The Price of a Good Paper," in "Harlow's Wooden Man." Special issue, *Quarterly Journal of the Marquette County Historical Society* (Fall 2008).

Roosevelt, W. Emlen. *Roosevelt vs. Newett*. New York: Privately Printed, 1914.

Shiras, George 3d. *Hunting Wildlife with Camera and Flashlight*. Washington, D.C.: National Geographic Society, 1935.

Wall Street Journal. "What Teddy Sipped." February 18, 2006.

Index

A

B

C

D

About the Author

Mikel B. Classen has been writing about northern Michigan in newspapers and magazines for over thirty-five years, creating feature articles about the life and culture of Michigan's north country. He's written about history, travel, the outdoors and many other subjects. A journalist,

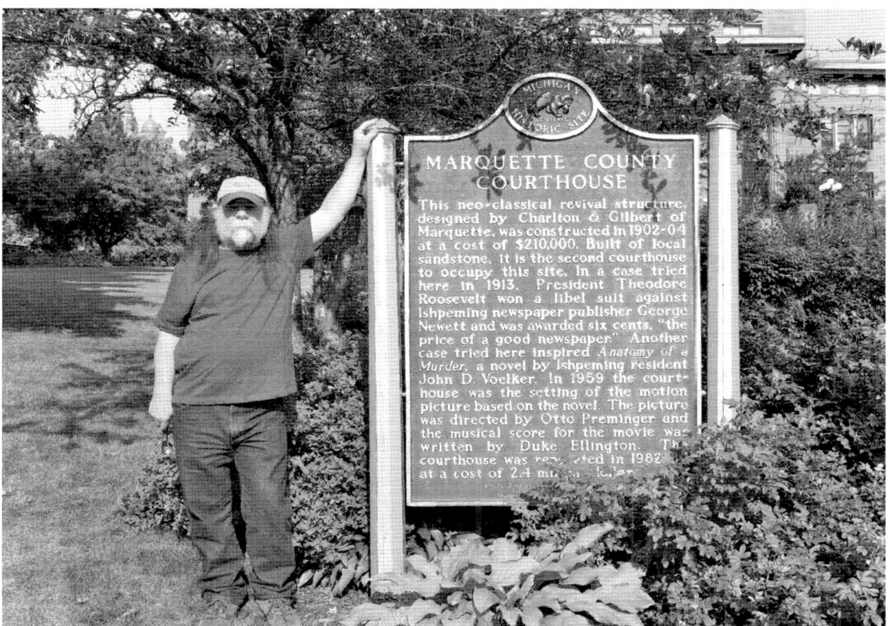

historian, photographer and author with a fascination of the world around him, he enjoys researching and writing about lost stories from the past.

Classen has a passion for history and makes his home in the oldest city in Michigan, historic Sault Ste. Marie. He is also a collector of out-of-print history books, historical photographs and prints of Upper Michigan. At Northern Michigan University, he studied English, history, journalism and photography. He lives with his wife, Mary L. Underwood, and his Labrador retriever, Grand Sable Dune.

To learn more about Mikel B. Classen, check out his website at www.mikelclassen.com.